湖州地方特色文化英译丛书

湖州茶香飘千年
——湖州茶文化史纲

Tea History of Thousands of Years: An Outline of Huzhou Tea Culture

张西廷 主编

孙庆文 陈继红 段自力
游玉祥 刘雅峰 朱长河

图书在版编目(CIP)数据

湖州茶香飘千年 / 湖州茶文化史纲 = Tea History of Thousands of Years: An Outline of Huzhou Tea Culture : 英文 / 张西廷主编 ; 孙庆文等译. — 苏州 : 苏州大学出版社, 2019.9

(湖州地方特色文化英译丛书)

ISBN 978-7-5672-2712-5

Ⅰ.①湖… Ⅱ.①张… ②孙… Ⅲ.①茶文化-文化史-湖州-英文 Ⅳ.①TS971.21

中国版本图书馆 CIP 数据核字(2018)第 291466 号

书　　名: Huzhou Chaxiang Piao Qiannian——Huzhou Chawenhua Shigang
湖州茶香飘千年——湖州茶文化史纲
Tea History of Thousands of Years: An Outline of Huzhou Tea Culture

译　　者: 孙庆文　陈继红　等
责任编辑: 杨　华
装帧设计: 刘　俊

出版发行: 苏州大学出版社(Soochow University Press)
社　　址: 苏州市十梓街 1 号　邮编: 215006
印　　刷: 虎彩印艺股份有限公司
网　　址: www.sudapress.com
E - mail: sdcbs@suda.edu.cn
邮购热线: 0512-67480030
销售热线: 0512-67481020

开　　本: 700 mm×1 000 mm　印张: 14.25　字数: 272 千
版　　次: 2019 年 9 月第 1 版
印　　次: 2019 年 9 月第 1 次印刷
书　　号: ISBN 978-7-5672-2712-5
定　　价: 48.00 元

Preface 1

Huzhou, also named Wuxing in ancient times, is located in the center of the Yangtze River Delta, which is the most developed region of China. It is an ancient city of rivers and lakes in the south of the Yangtze River, lies to the south of the wavy Taihu Lake, and is the only city named after Taihu Lake, which is 800-*li* wide and covered with mist and waves. It has been nearly 2,300 years old since 248 BC, when Huang Xie, also called Lord Chunshen, one of the Four Well-known Lords of the Warring States Period, brought civilization to the Eastern Tiaoxi River, by building Gucheng Town, setting up Gucheng County, and opening up farmland and water conservancy along the river.

Huzhou is not only an ancient city of history but also a city of culture. For thousands of years, diligent people here have smartly created a variety of brilliant cultures, such as silk and sericulture, culture of Huzhou writing brush, painting and calligraphy, tea culture, bamboo culture, fishing culture, food culture and so on. All these local cultures have made the beautiful and prosperous land to the south of Taihu Lake a famous "home of silk, habitat of fish and grain, and land of culture" of China and have contributed a great deal to civilization and prosperity of Chinese nationality.

The history and culture that our ancestors created and left us are valuable cultural wealth and indispensable spiritual resource in the development of Huzhou. Ever since the reform and opening-up of China, Huzhou's economic strength has obviously been enhanced, people's living standards have significantly been improved, the achievements have been huge, and the city has become more and more beautiful day by day. All these great changes above are attributed to the leadership of our Party, the great practice of people, and inheritance and innovation of excellent history and culture.

Culture is the support of spirit, the power of progress, and the assurance of harmony. Huzhou is now in a great era of development in that it aims to be a modern and ecotypic city and takes a lead to rise up from the development zone of Hangzhou, Huzhou and Nanjing of the Yangtze River Delta. To realize this grand objective, we should keep on enhancing our strength, stimulating our vitality, and revealing our charm; we should greatly develop the advanced culture and we should further inherit, carry forward and innovate its excellent historical culture.

It is the first time that *The Series of the History and Culture of Huzhou* has systematically revealed the historical civilizations of thousands of years, eulogized a great many outstanding talents in history and depicted the peculiar social morals. The publication of this series is not only a great fruit of the culture construction but also a significant medium to inherit and carry forward the excellent and traditional cultures. Basing ourselves on the reality and standing at the height of new times, we are to create a new history and civilization in the course of modernization construction of Huzhou on the basis of history and culture of ancient times. Only in this way, can it be worthy of our ancestors, the best inheritance and innovation of its history and culture.

New attempts of research have been made in composing this series to cover the length and variety of history and culture of Huzhou. However, we are sure to have some imperfections due to the limitation of time and sincerely welcome suggestions from readers. In the meanwhile, more researchers are hoped to join in the research of the history and culture of Huzhou so as to make the excellent ancient civilizations shine today.

The Secretary of Huzhou Municipal Committee of the CPC
The Director of Standing Committee of Huzhou Municipal People's Congress
Sun Wenyou
April, 2007

Preface 2

From the threads a mother's hand weaves,
A gown for parting son is made.
Sewn stitch by stitch before he leaves,
For fear his return be delayed.
Such kindness as young grass receives,
From the warm sun can't be repaid.

(Translated by Xu Yuanchong)

The poem is written by Meng Jiao, poet of the Tang Dynasty. Meng Jiao is a native of Huzhou and these immortal lines of his often take me back home—Huzhou. It is indeed much too difficult for me to express my homesickness actually. Though I was born in Wuhu of Anhui Province, I vaguely remember my childhood. What I could remember is Wuxing (now Huzhou) of Zhejiang Province which was the native place my family usually report. My eldest brother, being the eldest branch grandson, has so far kept a copy of Family Tree of Zhang of Dixi of Wuxing. It is this ancient family tree that makes me, the wanderer outside my hometown since childhood, get to know something about the origin of my family.

In 2000, invited by Huzhou Municipal Committee of the CPC and Huzhou Municipal Government, I was back home to attend Huzhou Culture Festival, whose theme is "Paintings and Calligraphy in the South of the Yangtze River", and delivered a speech at the Huzhou Writing Brush Forum. Since then, I have been back to Huzhou many times thanks to the kindness from leaders and fellow people of my hometown. My family belongs to Qingfentang Branch of Zhang Family of Dixi of Huzhou. Though having been outside for generations, we have been clearly conscious that we are deeply rooted in Huzhou. In my memory, every spring before the Chinese People's War of Resistance against Japanese Aggression, representatives of my family must be sent to Digang and Hangzhou to visit our ancestors' graves, and the grave keepers always sent us specialty of Huzhou such as bamboo shoots and lotus roots through others. The children of the family all use Huzhou writing brushes to practice writing Chinese characters and to imitate handwriting of Zhao Menghu. Up to now, we have been celebrating our holidays and the Spring Festival according to the traditional custom of hometown. Every time I return to hometown with deep nostalgia, I have always been warmly welcomed by the leaders of Huzhou Municipal Government and the fellow townsmen of Digang. My wandering feelings completely disappear in the mountains and waters as well as the simple and amorous affectations. I am much too pleased to witness the great changes of Huzhou and Digang and every progress in the economic development. Life is beautiful and the world is colorful, but my greatest desire lies in returning to hometown just like falling leaves finally returning to the root. Whether I deliver speeches at home or abroad, homesickness can always make me spiritually comfortable.

"The flowers come out at the same time annually, but the people have changed differently." Being a wanderer, I am greatly impressed and much pleased by the great annual changes and progress in Huzhou. Time flies and just before the Third Writing Brush Festival of Huzhou I am invited to check and approve a draft of *The Series of the History and Culture of Huzhou*. Having finished checking the whole set, I arrive at the conclusion that the series on the whole shows us the colorful and shining achievements of the history and culture. These achievements range from ancient times to modern times.

Huzhou centralizes the essence of sun and moon and has attractive mountains and waters. Ever since ancient times, the beautiful landscape of Huzhou has been well-known to the whole world, for its beauty extends from Qianshan Pond to Bilang Lake, from the Eastern and Western Tiaoxi River by Tianmu Mountain to Taihu Lake which is 800-*li* in width. Huzhou looks like a painting with mountains, rivers and plains interwoven with one another. It has a chain of mountains ranging out of sight, variegated rivers and charming plains. It has plenty of resources and the people here are living prosperously, comfortably, and honestly. Since the Tang and Song Dynasties, it has become popular to read books and compose poems by drinking. Since ancient times, it boasts beautiful mountains and rivers and has talents coming out in succession. From ancient times to now, countless celebrities and men of letters have great to do with Huzhou; namely, they are Wang Xizhi, Chen Baxian, Li Bai, Yan Zhenqing, Shen Yue, Zhang Zhihe, Jiaoran, Lu Yu, Meng Jiao, Su Dongpo, Zhao Mengfu and Wu Cheng'en. They have all overcome great difficulty to come here and left their traces, poems and works. They have made indelible contributions to Huzhou, whether they are natives here or not.

As the saying goes, people are known for the place they live in and the place becomes known for its peculiar inhabitants; people are known for the literature they composed and the literature become known for its peculiar inhabitants. In the 21st century, Huzhou is now in a new era and its long history and culture have been valued and cherished. *The Series of the History and Culture of Huzhou* presents us with the cultural achievements ranging from ancient times to the present day, the natural resources and customs, the thousand-year deposits of regional civilization of "the habitat of fish and grain and the home of silk". Zhang Jianzhi, one of my cross-age friends, has made many calls to Professor Peng Xiangnan, one of my students, to ask me to write a preface for the series. However, I have postponed writing it till now, because at that time I was in Japan for an academic purpose and was fully engaged. Now I take this opportunity to inform my fellow townsmen that I am communicating with Daisaku Ikeda, a famous Japanese learner, and doing my best to promote cultural exchanges between China and Japan. I, being in Japan, would like to express my heartfelt congratulations on the publication of *The Series of the History and Culture of Huzhou.* I hope that this series will provide the readers with spiritual wealth and will enrich their minds so as to attract more people to know and love Huzhou as it is and as it was.

I kindly wish Huzhou a more beautiful future. This is my preface for *The Series of the History and Culture of Huzhou* and I take it as my return to Huzhou, my hometown.

Tenured full professor, former president of Central China Normal University

Zhang Kaiyuan

in Tokyo

April 18th, 2007

Contents

Chapter 1

Sprouting Tender Bud:

Huzhou Tea Culture in the Qin and Han Dynasties

1.1 Remarks before the Sprout of Tea Culture of Huzhou

1.1.1 International Academic Views on the Origin of Tea

As early as in the Tang Dynasty, Lu Yu wrote in *The Classic of Tea*, "Tea trees, the best wood of the South, can grow to one foot, two feet, and even dozens of feet tall. In the canyon of Sichuan, there are even tea trees of the folds of two men which can only be cut down to pick tea leaves." His above vivid description tells us that in the middle of the Tang Dynasty more than 1,200 years ago, there were tea trees which grew as tall as ten feet and in the folds of two persons. In 1753, the Swedish naturalist Carl von Linné named the tea as *Thea sinensis* (Chinese tea trees). Therefore, in the early times, there was no objection that tea originates in China. However, in 1824, British scholar Bruce discovered some tall tea trees in the Indian state of Assam and concluded that these tea trees, more than 10 meters high, were the most ancient wild ones in the world. Later in 1838 in London, he said in his book that it was in India instead of in China that the tea trees originated. At that time, though the territory of China was rich in tea leaves, there was no discovery of wild ancient tea trees. In the next 100 years, the British scholars maintained that the world tea trees originated in India. During this period, ancient tea trees were also found in Mauritius in Africa, which could also be called the origin place of tea trees. Even though there had already been an agreement that tea trees originated in China, continuous arguments on the origin place have been on ever since the 1820s. As to the origin place of tea trees, there were four different opinions: a) China; b) India; c) two origins: broad-leafed tea in India and small-leafed tea in China; d) multi-origins: all the natural areas which were fit for growth of tea trees were the origin places of tea trees.

In October, 2005, *Xinhua Daily News* published a series of reports of "Ten-thousand-*li* Exploring of Ancient Tea Horse" by a party of five reporters such as Wu Hao and Liu Juan from Xinhua News Agency. The titles of these reports were as follows: "The Mysterious Tea Horse Road in Modern Times" "Tea Trees of 35.4 Million Years by the Ancient Tea Horse Road" "Simao: The Place where Tea Trees Survived Quaternary Glacier" "Simao, the Largest Area of Ancient Tea Trees in the World" and "The Latest Research: Simao of China is the Birthplace of the World Tea Trees". These series of reports are comprehensively arranged as follows.

In 1978, the fossils of broad-leafed magnolia (new species) were discovered in Mangxian, Jinggu Basin, Yunnan Province, China by the scientists from the Institute of Botany of the Chinese Academy of Sciences and Nanjing Institute of Geology and Paleontology of the Chinese Academy of Sciences. These magnolia fossils were about 35.4 million years old. The magnolia is the source of angiosperm and the ancestor of vertical evolution of camellia. The fossils of broad-leafed magnolia unearthed in Mangxian, Jinggu, eloquently prove that Simao-centered area in the southwest of Yunnan Province, China is the most primitive origin of the tea trees. Simao City, Yunnan Province, China becomes the only location to preserve the complete chain of vertical evolution of tea plants in that there are broad-leafed magnolia fossils of 35.4 million years old in Mangxian of Jinggu, ancient wild tea trees of 2,700 years old in Qianjiazhai of Zhenyuan, a transitive type tea trees of more than 1,000 years old in Bangwei Village of Lancang, and a cultural king of tea trees of more than 1,000 years old in Jingmai of Lancang.

According to Darwin's theory of Origin of Species, scientists come to the agreement that the judgment of the origin place of a species comes from three necessary factors: in the first place, the original biological characteristics of the species; in the second place, the complete vertical evolution system of the species; finally, the geographical distribution of the most primitive flora of the species. All the three factors are indispensable. India cannot be the origin place of a species for only a few wild ancient tea trees have so far been found there. Up to now, Simao in Yunnan, China has been the only location to have all these three factors available, and it can be called the only origin place of a species of tea trees in the world.

The tea trees and ancient magnolia fossils have the genetic relationship between them for they share more similarities. The tall wild tea trees found in western Yunnan may have evolved from broad-leafed magnolia of the Tertiary Period of the region by *Magnolia miocenicas.* Meanwhile, free from the destruction in pleistocene of the Quaternary Period, these wild tea trees have survived, reproduced and developed to a large scale within the territory of Simao, Yunnan. After the preliminary survey and confirmation, ancient tea trees cover an area of 2.4 square kilometers in Simao, and the tea tree species have amounted to 31 species and 3 variants of 4 series. And the world record of the types of tea trees is only 37 species and 13 variants of 4 series. Simao of Yunnan Province, China has become the location for plant communities with the ancient wild tea trees as the dominant species in that the area is the largest, the types are the richest, oldest and the most complete. In 1993, 181 paleontologists and tea experts from 9 countries and areas got together in Simao to testify and demonstrate a series of findings on ancient tea trees and ancient tea gardens. Through their field demonstration, they have officially confirmed that Simao City, Yunnan Province, China is the world center of the origin of tea trees. The case of the origin of tea trees has finally been settled after a century's dispute.

1.1.2 A Long History of Tea and Tea Drinking in China

China has a long history of tea and tea drinking. Nowadays, even in the scene of luxury and noisy agitation, some people can always be fond of making a cup of tea, and the fragrance over the tea will give them peace and contentment.

China is the earliest country to discover and utilize the tea for its tea history is thousands of years long. Tea actually originates in the southwest of China, and tall wild tea trees of more than 1,000 years old can still be found in Yunnan and other places. According to the historical record, the ancient Bashu region, in the vicinity of Sichuan and Hubei Provinces today, is one of the birthplaces of Chinese tea culture. From the Tang and Song Dynasties to the Yuan, Ming and Qing Dynasties, Chinese tea production area has been continuously expanded, and tea culture has continuously developed and gradually spread to all over the world. Chinese tea, an ancient drink, has made a positive contribution to civilization and progress of mankind.

It is recorded in Chapter Ⅵ "Tea Savoring" of *The Classic of Tea* by Lu Yu of the Tang Dynasty that "Tea drinking is discovered by Shennong". In his book *An Outline of History of Chinese*, Bo Yang states "Shennong is a god of the people in the age of mythology". Legend has it that Shennong is the ancestor who invented the primitive agriculture and medicine. In China, tea was discovered and used in the primitive and matriarchal society. Ever since the Shennong Era, it has been about five or six thousand years.

Legend says "Shennong was poisoned seventy-two times in the course of tasting wild herbs, and he found a solution in tea". Shennong is also called Yan Emperor, one of the three ancient emperors of China. Legend has it that before 2700 BC, Shennong always went into the wild mountains to gather herbs for medical purpose. Not only did he have to cover a long way, but also he collected and tried herbs personally to experience and identify the functions of every herb he got in hand. One day, Shennong was poisoned when tasting a wild herb and felt dry, numb, and dizzy. He quickly sat down by a tree to have a rest with his eyes closed. Just then, a gust of wind blew down a few green leaves from the tree. Shennong accidentally picked up two pieces and put them into his mouth, chewing. To his surprise, fresh fragrance had his tongue wet, and Shennong felt spirited. His sickness disappeared. He was very curious to pick up a few more leaves for careful observation. Then, he found them different from other leaves in the shape, vein and margin. He collected some leaves and took them back for careful study. Later, he named the leaves as "tea".

So tea drinking begins with medical use. Historical facts of ancient times are often covered with a myth in the ancient literature. We can only look for the footprint of ancestors in making history through the mythology.

Chang Qu, historian of the Eastern Jin Dynasty, recorded in his book *Records of Huayang* that about 1000 BC, when King Wu of the Zhou Dynasty fought against King Zhou of the Shang Dynasty, in Bashu region tea began to be used as a treasured "tribute", which is the earliest recorded tribute tea. When King Wu of the Zhou Dynasty beat King Zhou of the Shang Dynasty and eliminated the Shang Dynasty in 1066 BC, the tribe chiefs of Bashu region began to send "vermilion, lacquer, tea, and honey" to the Zhou Dynasty as tributes. The book also says that good tea came from Sichuan Mountains and the Western Zhou Dynasty had an official post called "Tea Manager" according to *Rites of the Zhou Dynasty*.

From the Spring and Autumn and the Warring States Periods to the Qin and Han Dynasties, the Three Kingdoms Period, the Western and Eastern Jin Dynasties, the Southern and Northern Dynasties, tea drinking began to be popular gradually. Although there was no tea monograph written then, a good many activities about tea and tea drinking appeared in poetry, prose, essays, critical interpretation of ancient texts, biographies, chronicles and local annals by famous personages. At the end of the Spring and Autumn Period, Confucius (551 BC-479 BC), famous thinker, statesman, educator and founder of Confucianism, repeatedly mentioned the "tea" in the censor of *The Book of Songs*. Yan Ying (? - 500 BC), Minister of the Qi State in the Spring and Autumn Period, once drank tea dishes. Sima Xiangru (178 BC-118 BC), poet of the Western Han Dynasty, wrote the tea names in his book *Fanjiang Pian*, a book on philology. Yang Xiong (53 BC-18 AD), litterateur, philosopher and linguist of the Western Han Dynasty, recorded tea activities in the southwest Shu Kingdom in his book *Dialects*. Wang Bao, litterateur of the Western Han Dynasty, wrote the words such as "having all the tea sets for making tea" and "to buy tea in Wuyang" in his essay *Tongyue*, an essay on common culture of the Han Dynasty. Hua Tuo (? - 208), medical expert of the Eastern Han Dynasty, related medication function of tea in his *On Food*. In the Three Kingdoms Period, there was a story about Sun Hao(242-283), Duke of Wucheng, drank tea instead of liquor. Liu Kun (271-318), litterateur of the Western Jin Dynasty, wrote to his nephew, who was an official in Yanzhou, to ask him to buy tea for him. Fu Xian (239-294), poet of the Western Jin Dynasty, told a story of an old Sichuan woman selling tea in his *Instructions from an Inquisitor*. Zhang Zai, litterateur of the Western Jin Dynasty, wrote a line of "Outshining all other kinds is Sichuan tea" in his poem "Stepping on Baitu Building in Chengdu". Du Yu, poet of the Western Jin Dynasty, wrote a book titled *Ode to Tea*. Zuo Si (250? - 305?), litterateur of the Western Jin Dynasty, described two charming daughters making tea in his poem "My Cute Daughters". Wang Xizhi (303 or 321-361 or 379), great calligrapher of the Eastern Jin Dynasty, had two Chinese characters *cha* and *tu* in his calligraphy. Guo Pu (276-324), litterateur of the Eastern Jin Dynasty, in writing notes for *Erya* (the earliest dictionary), said "Cha is called *tu* in the morning and *ming* in the evening". Gan Bao, historian of the Eastern Jin Dynasty and Tao Qian (365 or 367-427), litterateur of the Eastern Jin Dynasty, wrote stories of drinking tea in their books titled *Anecdotes about Spirits and Immortals* and *Sequel to Anecdotes about Spirits and Immortals*. Lu Na, Wuxing (now Huzhou City) Satrap of the Eastern Jin Dynasty, treated his guests with tea. Liu Yiqing (403-444), Duke of Linchuan of the Song Dynasty of the Southern Dynasty, recorded folk tea activities in his book *A New Account of Tales of the World*. Liu Jingshu had folk tea activities in his book *Weird Garden*. Shan Qianzhi of the Song Dynasty of the Southern Dynasty related imperial tea was produced in Wenshan Mountain of Wucheng County in *The Records of Wuxing*. Fayao, monk of the Southern Dynasty, drank tea for each meal when he was in Xiaoshan Temple of Wukang (now in Huzhou City). He was still healthy at the age of 80. Bao Linghui, litterateur and poetess of the Song Dynasty of the Southern Dynasty, composed her "Ode to the Fragrant Tea". Tao Hongjing (456-536), thinker of Daoism and medical expert of the Qi and Liang Dynasties of the Southern Dynasty, mentioned that tea could make one feel light and strong in his work *Miscellaneous Notes*. Wang Su (464-501) of the Southern and Northern Dynasties loved drinking tea and milk. One day he was asked, "Which is better, tea or milk?" He answered, "Tea is, by no means, above milk." Thus he began to call tea *laonu*. The allusion was used in poems and articles, and ever since then *laonu* has become a synonym of tea.

The above descriptions of tea and tea activities concerning historical celebrities have mostly been adapted from *Dictionary of Celebrities in Chinese History*, *Dictionary of Chinese Ancient Prose*, *Random Notes of Chinese Tea History*, *The Classic of Tea* and other works. All these records indicate that tea drinking became popular before the Tang Dynasty.

1.1.3 On the Legend of Fangfengshi and the Baked Bean Tea

In China and even in the world, Huzhou is one of the earliest places where tea is grown and utilized. Throughout the long history, there is not much detailed historical data passed down, but a good many folk stories, legends, and customs still offer us valuable clues for further study of tea. Out of all the stories and legends, one about a special custom tea must be mentioned, that is, Baked Bean Tea. This tea is also well-known as Magic Tea of Fangfengshi, which has been popular and available up to now in the vicinity of Erdu countryside of Deqing County, Huzhou City.

Baked Bean Tea is a salty tea, mainly made of baked cooked beans, dotted with bud tea and flavored by orange peels and wild sesame seeds. Baked Bean Tea has varieties of seasoning in different seasons. Generally, judgment is made when boiling water is first served; flavor is tasted when boiling water is served for the second time; after boiling water is served for the third time, people usually drink the water and eat all the baked beans, tea leaves and the seasoning. Baked Bean Tea is not only fragrant and delicious in taste, but also rich in nutrition. It is actually a traditional healthy beverage of all ages.

According to legend in ancient times, Fangfengshi was sent by Yu to Xiazhu Lake to control flood, working day and night. In reward to his hard work, the local people made tea seasoned with orange peels, wild sesame seeds and others for driving away dampness and making him refreshed from labor. People also baked beans for him to eat with tea. Once Fangfengshi accidentally poured baked beans into tea water. He then drank tea water and ate the beans. After that he was magically strong and powerful to put mountains into sea and the flood was finally controlled. From then on, Baked Bean Tea has been very famous and has been handed down from generation to generation for more than four thousand years. Baked Bean Tea was regarded the best in every dynasty, the best to entertain guests, and gifts to friends and relatives. Lu Yu, the Tea Sage, recorded this famous salty tea in *The Classic of Tea*, which was written by the Eastern Tiaoxi River, the hometown of Fangfeng.

Then questions come: What drink is Magic Tea of Fangfengshi? What function does it have? Is it scientific?

1.1.3.1 Magic Tea of Fangfeng Is Both Healthy and Herbal Tea

Due to the mild climate, Taihu Lake in ancient times had much rainfall and flood year after year. It was first frozen in 1111 AD. The ancestors in the south of the Yangtze River came to recognize that the tangerine peels, perilla seeds, and high mountain tea were of herbal function to cure cold and rheumatism and gradually formed the habits of brewing and drinking tea so as to prevent or cure diseases. In the Three Kingdoms Period, tea was recorded as herbal drinks in history, namely Orange Tea, Sesame Tea, Ginger Tea, Mint Tea and so on. In the Yuan Dynasty, there appeared the tea diet. Chinese Wolfberry Tea, and Yumo Tea (mixed with fried rice and Zisun Tea) were recorded in *Principles on Proper Diet* written by Hu Sihui. In the Ming Dynasty, there were six kinds of tea soups recorded in *Book of Daily Usages*, of which Noon Tea had been developed for the treatment of cold. Magic Tea of Fangfengshi was characterized by simple brewing, choice of wild materials and best flavor, and it has been passed down up to now. A cup of Magic Tea of Fangfengshi can make one feel very comfortable especially after one eats meat or greasy food.

1.1.3.2 Magic Tea of Fangfengshi—the Most Primitive Ritual Drink in the South of the Yangtze River

In ancient times, liquor was generally used to offer sacrifices to ancestors and spirits. China's liquor brewing and drinking customs began in the Shang Dynasty, when liquor was used to offer sacrifices to ancestors. As early as four thousand years ago, people in the ancient Fangfeng Kingdom began to use Flavor Tea in ritual services for ancestors and spirits. After Fangfengshi was wrongly killed, the local inhabitants specially used this Flavor Tea to offer sacrifices to their ancestors and Fangfengshi.

It was very popular to use tea to worship ancestors and spirits in the Tang Dynasty. Monks used tea as the main worship drink for Buddhism advocated abstinence from drinking. Monks in temples of Zen Buddhism in the Tang and Song Dynasties were the most particular about tea. Qian Yi, a native of Lin'an (now Hangzhou City) recorded in his work *A New Book of the Anecdotal Stories of the Tang Dynasty* that, in 849 AD, an eminent monk in the eastern capital of Luoyang was 120 years old. Emperor Xuanzong asked him what drug could make him live long. He replied, "It is nothing but that I am so fond of tea that it is the first thing to get wherever I am." At the same time, lots of historical data shows that many kinds of China's most famous tea have been cultivated and processed in the Buddhist and Taoist temples. At that time, there was "Chief Manager of Tea" in the Buddhist temples to manage tea activities. At the temple gate, there were "tea-serving monks". The best tea was offered to the Buddha, the second best for guests and the inferior for self-service. Common people generally used tea to worship ancestors and receive guests. It can be said that Magic Tea of Fangfengshi is the most primitive ritual drink in the region of the south of the Yangtze River. It is also the longest tea custom in Chinese tea culture.

1.1.3.3 Magic Tea of Fangfengshi, the Powerful Evidence for the Ancient Fangfeng Kingdom

As early as four thousand years ago, the ancient Fangfeng Kingdom, located in China's Qiantang River and the Taihu Lake basin, covered the area of present Huzhou City and its Deqing, Changxing and Anji counties, Wujiang County (now Wujiang District of Suzhou City) of Jiangsu Province and towns such as Penggong, Pingyao, and Liangzhu of Yuhang City of Zhejiang Province. The capital of the ancient Fangfeng Kingdom was constructed at Fangfeng Hill of Erdu of Sanhe Town, Deqing County, Huzhou City now. The kingdom was at the end of patriarchal clan society. In agriculture, canals were dug to drain flood, wells were dug for drinking water, rice and peanuts were planted, and mulberry trees were planted to feed silkworms. In the handicraft industry there were Liangzhu black pottery, which was as thin as egg shells, and a variety of carved jades and horns. In water transportation, there were oars and boats. In culture, the earliest written words might have appeared. This was the period when the private ownership system appeared, and the primitive society was transiting to the slave society. In 2198 BC, King Yu, chief of military alliance of Huaxia Tribe, called local princes to a meeting at present-day Kuaiji Mountain, Shaoxing, in his patrol in the south of the Yangtze River. Because Fangfengshi once tried to dissuade and went against the attempts of King Yu to undermine the primitive democratic abdication system to pass down his throne to his son Qi, King Yu persecuted Fangfengshi with the excuse of his being late for the meeting. This is the first and biggest injustice in the history of our country. The ancestors of the Fangfeng

Kingdom also suffered a lot and they went exile, and part of the Fangfeng ancestors even went across sea to the islands of Japan.

There are such records in *General History of China · Prehistory*, *Records of the History · Records of Xia*, *The Outline of History of Chinese People* and *Dictionary of Celebrities in Chinese History*. All these works mentioned above had the following records: Yu, whose real name was Si Wenming, was also called Xia Yu, because he was conferred as Earl of Xia. Yu was the son of Gun, and Gun was the son of Zhuanxu. Zhuanxu was the son of Changyi, and Changyi was the son of the Yellow Emperor. In a word, Yu was the great-great-grandson of the Yellow Emperor. He was the leader of Xiahoushi Tribe of descendants of Xia and took orders from Shun to control floods, to conquer Sanmiao Tribe, to do patrols in the south and to hunt in the east, to summon all the dukes to a meeting, and to divide the kingdom into nine parts. Therefore, Yu made the greatest contribution to his country. After Shun's death, Yu was made the head of the Tribal League, and became the first ruler in the history of China. In order to get further integration and unification of all tribes, Yu first went to Tushan to do patrols in the south and to summon all the dukes to a meeting. Later he went to Kuaiji to summon all the Dukes of China to a meeting. At that time, the chief of Fangfengshi Tribe, King of Wangwangshi, was guarding Fengshan Mountain and Yushan Mountain (now in the territory of Wukang, Deqing County, Huzhou City) failed to arrive on time for the meeting, and King Yu killed him for his neglecting his duty and made it an example to all the dukes.

Records of the History · Confucian Family records a conversation by Confucius and an emissary of the Wu Kingdom about Fangfengshi.

The Wu Kingdom came to conquer the Yue Kingdom, and destroyed Kuaiji, the capital of the Yue Kingdom and got a bone, whose length could fill a carriage. The Wu Kingdom sent an emissary to ask Confucius, "Whose bone is the biggest?" Confucius answered, "King Yu summons all the dukes to a meeting at Kuaiji Mountain. Fangfengshi, one of the dukes, fails to come on time and is killed by King Yu. His dead body is left lying on the street as a warning to others. A bone of his can fill a carriage. This is the biggest bone." The emissary of the Wu Kingdom then asked, "Who is God?" Confucius answered, "The spirits of the mountains and rivers, who can make clouds and rains to benefit the world, safeguard mountains and rivers, who are always on time to worship spirits, are called Gods. Those who defend countries are called dukes. All the Gods and dukes are governed by the king." The emissary, then, asked, "What is the duty of Fangfengshi?" Confucius answered, "The chief of Wangwangshi guarding Fengshan Mountain and Yushan Mountain, and his family name is Li. The chief is called Wangwangshi in the Dynasties of Yu, Xia and Shang, called Changdi in the Zhou Dynasty, and Longman nowadays." The emissary then asked, "How tall are they?" Confucius answered, "Jiaoyaoshi is as tall as three *chi* and is the shortest, and the tallest one is no taller than three *zhang*." At this the emissary of the Wu Kingdom exclaimed, "What a great saint he is!"

Shen Kuo, scientist and statesman of the Northern Song Dynasty, had a story in his great book *Brush Talks from Dream Brook.* The story goes like this "One day someone discovered a rotten pestle for pounding silk. He did not know what it was, so he took it home and showed it to his neighbors. All his neighbors took a close study at it but they all were shocked for no one knew what it was. Later, a scholar passed by and studied it and concluded, 'This is holy. I have heard Fangfengshi is as tall as about three *zhang*. A bone of his can fill a carriage. This is a tibia of his.' The local people were very happy to hear that and they built a temple to worship this bone." *Records of Wukang County* written in the Qing Dynasty records: "Wangwangshi's castle lies in the southwest of Wukang and is 15 *li* away. It is also called the Giant Castle where the descendants of Fangfengshi live and survive."

In the years of Daoguang Period of the Qing Dynasty, *Records of Wukang County* records: "Between Fengshan Mountain and Yushan Mountain lies the ancient kingdom of Wangwangshi, which is Fangfengshi. A temple for worshiping Fangfengshi is built at the foot of the mountain. It is built by He Xun, County Magistrate, at the beginning of Yuankang Period of the Western Jin Dynasty. Qian Liu, Emperor of the Wuyue Kingdom, often comes here to make his prayers and finds they are efficacious and confers him the title of King Lingde, and the stelae are still there. In 1371, Hongwu, Emperor of the Ming Dynasty, confers the title of God of Fangfengshi. Up to now, there has been still a service of worshiping Fangfengshi God with sheep, wine and baked beans in every lunar August 25." The history records that the people of the ancient Fangfeng Kingdom worship

Fangfengshi with "sheep" "wine" and "beans" (Baked Bean Tea), and the custom has been passed down.

Erdu, the capital of the ancient Kingdom of Fangfeng, was closely connected with Liangzhu Culture and Qianshan Pond Relics, and especially with the triangle area of Liangzhu, Pingyao, and Anxi having been excavated in recent years reflecting the ancient Kingdom of Fangfeng four thousand years ago. The area of Hangzhou and the area in the south of Liangzhu were gulf, lakes or marshes. Fengshan Mountain in Erdu of Deqing (formerly called Fangfeng Mountain), Yushan Mountain (formerly called Zhangzi Mountain) and Xiazhu Lake (formerly called Yazi Lake or Fangfeng Lake) were the most beautiful at that time.

Since the beginning of the Spring and Autumn Period, Mogan Mountain of the ancient Fangfeng Kingdom has become a famous mountain for ore mining and melting. Lishan Mountain and Xishizhen of Liuqiao Village beside Ganshan Mountain to the southeast of Erdu had become the secluded place for Fan Li, a senior official of the Yue Kingdom, and Xishi, the ancient beauty of China. With the development of Taoism and the introduction of Buddhism, a great many temples had been built in the Fangfeng Kingdom and its surrounding areas. These temples attracted celebrities and famous monks of China, and some even chose to live there. For example, attracted by Chinese culture, Monk Kang Senghui (?-280) came a long way from Sindhu (now India) to Jianye (today Nanjing City) in 247 to build temples there. Later, he built temples at Jinsu Mountain (now in Haiyan County) and he became the earliest Buddhist leader in the south of China. In 291, the first year of Yuankang Period of the Western Jin Dynasty, He Xun, County Magistrate of Wukang, built the Temple for Fangfengshi in the south of Fengshan Mountain (today Fangfeng Mountain). In 931, Qian Liu, King of the Wuyue Kingdom, gave the title of King Lingde to Fangfengshi and enlarged the original temple into a bigger one titled as King Lingde Temple. Later, it was said that Baozhang (between 414 BC and 657 AD), an Indian Zen master, also called as Qiansui Monk, came all the way to China in 259 after he heard that the area in the south of the Yangtze River was a place of wonder. He visited all the well-known mountains and rivers in the south of the Yangtze River and built a hut to settle down at Zhongtianzhu of Hangzhou to the southeast of Erdu in 597 when he was 1,011 years old. He lived there for as long as 45 years and became the founder of Zhongtianzhu Temple. Baoen Temple in Shangbaibu, which was over 10 *li* away in the west of Erdu, became the most famous temple of the Tang and Song Dynasties. Up to the Ming and Qing Dynasties had emerged a batch of masterly Buddhist monks. Those temples all advocated tea ceremonies for they were influenced by tea culture of the ancient Fangfeng Kingdom.

1.2 Huzhou Tea Culture Beginning to Sprout

1.2.1 Huzhou Tea and Tea Culture in the Western and Eastern Han Dynasties

In the period before the Qin Dynasty, the production and utilization of tea mainly occurred in Bashu region. Since the Han Dynasty, tea production and drinking had extended gradually from Bashu region to all over the country, and had gradually spread to the vast area of the middle and low reaches of the Yangtze River in particular. The earliest record of tea in Huzhou was in the Han Dynasty. There were descriptions in *Appraisal on Dongshan Jiecha Tea* by Zhou Gaoqi of the Ming Dynasty that "there was an ancient Chinese emperor of the Han Dynasty who lived in the south of Mingling Mountain and taught the youngsters about tea". Mingling Mountain, which is over 500 meters above sea level, lies at the junction of Yixing City of Jiangsu Province and Baixian Town of Changxing County, Huzhou City. At the foot of Mingling Mountain lies Luojie of Baixian Town, and at the east of it lies Dongshanjie. Luojie is famous for its tea production and it is as famous as Guzhu which is famous for Zisun Tea (a kind of tea shaped like young bamboo shoots). Luojie is also the secluded place for the poet Lu Tong of the Tang Dynasty. The story of "teaching the youngsters about tea" tells us two things. One is that an ancient Chinese emperor of the Han Dynasty hired men to plant tea in the south of Mingling Mountain, and the other is that at that time the youngsters were taught about tea technology, which shows us that tea activities had developed to a certain scale. Although this is only a folk story, the relics excavated in Huzhou are enough to prove that Huzhou had produced tea and tea drinking in the Han Dynasty. In the early 1990s, in Huzhou was unearthed a celadon tea-storage urn of the Han Dynasty, unbroken and intact. Also known as four-ring linear pottery jar (presently collected in Huzhou City Museum), it is 33.5 cm high, with a maximum belly diameter of 34.5 centimeters, shoulder hunched and belly bulging. It has a flat bottom and concave inside, the four cross hangers lying symmetrically on the part of the shoulder. The hangers are surfaced with leaf-shaped lines, and the main body is decorated with chord-marked lines. The seal case has diamond-checked patterns. On the upper part of the urn's shoulder is engraved a Chinese character of "荼", meaning "tea", indicating a utensil for tea storage. The overall shape is complete, despite a slight crack in its body. The bottom is not glazed and the dripping trace of grazing can still be clearly seen with a greenish brown color. Unlike the common urn, it is glazed inside as well. This has been by far the earliest special tea ware inscribed with the Chinese character "荼".

1.2.2 Huzhou Tea and Tea Activities during the Three Kingdoms Period, the Western and Eastern Jin Dynasties and the Southern and Northern Dynasties

During the Three Kingdoms Period, the Western and Eastern Jin Dynasties and the Southern and Northern Dynasties, tea and tea culture in Jingchu region developed gradually and spread across the country. Due to the geographical advantages, the middle and lower reaches of the Yangtze River or the middle of China had become more important and taken the place of Bashu region in the spread of Chinese tea culture.

During the Three Kingdoms Period, the Wu Kingdom once occupied part of now Jiangsu, Anhui, Jiangxi, Hubei, Hunan, Guangxi and half of the southeast Guangdong, Fujian, and Zhejiang. This region was also a main tea-producing area in China's tea industry at that time.

Lu Yu recorded all the popular romance stories in *The Classic of Tea* such as Sun Hao, Duke of Wucheng of the Wu Kingdom, used tea in place of wine. Lu Na, Wuxing Satrap of the Eastern Jin Dynasty, used tea to entertain guests. Monk Fayao of Xiaoshan Temple of Wukang of the Southern and Northern Dynasties enjoyed a long life due to tea drinking. Wenshan Imperial Tea was produced in Wucheng of Huzhou during the Three Kingdoms Period. All the facts above prove that the production and consumption of tea in Huzhou were very popular in the Three Kingdoms Period and the Southern and Northern Dynasties.

1.2.2.1 Sun Hao, Duke of Wucheng of the Wu Kingdom of the Three Kingdoms Period, Used Tea in Place of Wine

Lu Yu cited the romance story of Sun Hao, Duke of Wucheng in the Three Kingdoms Period, from *The Chorography of the Wu Kingdom · Biography of Wei Yao* in the seventh chapter of *The Classic of Tea.* The story was like this: "Whenever Sun Hao holds a feast, all attendants must drink at least seven litters. When Sun Hao knows that Wei Yao can drink at most 2 litters, he allows Wei Yao to drink tea instead of wine." Feng Yan of the Tang Dynasty recorded in Volume Ⅵ of *Observations of Feng Yan*: "Whenever Sun Hao, King of the Wu Kingdom, gives feast to all his ministers, everybody must be drunk. Wei Zhao, who changed his name from Yao because of taboo of the Jin Dynasty, cannot drink much, so Sun Hao tells him in secret to drink tea in place of wine, which was recorded in *The Chorography of the Three Kingdoms* by Chen Shou, historian of the Western Jin Dynasty."

Sun Hao (242-284), styled name Yuanzong, also called Pengzu, was the grandson of Sun Quan and the son of Sun He. Sun Xiu, Emperor Jing of the Wu Kingdom of the Three Kingdoms Period, in 258 conferred Duke of Wucheng on Sun Hao, his brother's son (Wucheng County of the Three Kingdoms Period having jurisdiction over today's Wuxing, Nanxun, and Changxing County of Huzhou City). In the first year (264) of Yuanxing Period, Sun Hao succeeded to the crown and became the last emperor of the Wu Kingdom. *The Chorography of the Three Kingdoms · Wu* by Chen Shou recorded: "When Sun Xiu died, the Shu Kingdom had just been conquered by the Wei Kingdom. The whole kingdom is shocked at the rebels. Wan Yu, Left Captain of the Wu Kingdom, having once been Magistrate of Wucheng County, has very good relations with Sun Hao and often speaks highly of him as talented and decisive just like Sun Ce. Later on, he proposed to Sun Xiu's wife to take Sun Hao to be her son. When Dowager Zhu, Sun Xiu's wife, heard the proposal to take Sun Hao as her son, she said: "I, being a widow, have no worries over the kingdom. If the Wu Kingdom remains intact and we have the ancestral temple to worship, I agree to take him as my son." Hence Sun Hao became Emperor of the Wu Kingdom at the age of 23.

The first volume of *Annals of Huzhou Prefecture* written in Wanli Period of the Ming Dynasty recorded Wuxing County began to be called thus, which indicated that Sun Hao was a very talented successor of the Wu Kingdom. *The Records of Wuxing* by Shan Qianzhi of the Song Dynasty of the Southern Dynasty recorded all about geographic, natural resources and other conditions of the nine counties of Wuxing (referring to *The Great Dictionary of Chinese Local Chronicles*, published in 1988 by Zhejiang People's Publishing House).

Lu Yu cited *The Records of Wuxing* by Shan Qianzhi in Chapter Ⅶ of *The Classic of Tea* that Imperial Tea was produced 20 *li* away from Wenshan Mountain in the west of Wucheng County. Shan Qianzhi, historian of the Song Dynasty of the Southern Dynasty, was the descendent of Shan Tao, one of Seven Sages of Bamboos. In 443, the twentieth year of Yuanjia Period, he was a scholar of history and was appointed as the Magistrate of Jiyang County (now in the south of Nanyang of Henan Province). In the early years of Xiaojian Period, he took orders to compile a book of history, and died of unknown illness (referring to *Zhejiang Literature Series · The Sources of Compiling Chronicles of Zhejiang Province before the Ming Dynasty*). Sun Hao was the last emperor of the Wu Kingdom of the Three Kingdoms Period. The tea he drank in place of wine was Imperial Tea from Wenshan Mountain of Wucheng County, which was recorded by Shan Qianzhi in his book *The Records of Wuxing*. According to *Imperial Tea Produced from Wenshan Mountain* written by Lin Shengyou in 1986, "Wenshan, also called Nanyunfeng Summit, stands 504.9 meters high above sea level and is the second summit of Bianshan Mountains. It faces south with the summit of Bianshan Mountains in its northwest to protect it from the cold and dry airflow blowing southeast in winter. Tea trees grow in the deep valleys of Wenshan Mountain. It has great valleys, such as the Great and Second Valleys, Stone Mortar Shaped Valley, Yangjia Valley, and so on. All the valleys are collectively called as Wenshan Valley. This valley runs 3.5 kilometers long and has hot springs everywhere. In winter, the surrounding mountains are covered with white snow. However, in Wenshan Valley, the steam from the hot springs forms great fog which fills the air. The green trees are growing around the hot springs, which takes on a peculiar scene. This is the excellent environment for Imperial Tea of Wenshan, the historically famous tea, to grow there. "Although Sun Hao, the last emperor of the Wu Kingdom of the Three Kingdoms Period, had very bad fame in history, who was described as "tyrannical and dissolute" in that Sun Hao killed Wei Yao because he refused to obey his order to write a biography of Sun Hao's father, he actually did a great deal for the people there when he was Duke of Wucheng in Wucheng County. For example, Sun Hao built the Gaotang Canal in the west and the Suntang Canal in the south (referring to the 19th volume of *Annals of Wuxing* in Jiatai Period of the Southern Song Dynasty and *Wuxing Prefecture and the Culture of Distinguished Families*). In the history of China, Sun Hao is the first to use tea in place of wine, and the custom has been passed down as a clean and elegant one. These facts have proved that as early as in the Three Kingdoms Period, tea and tea drinking began in Huzhou.

1.2.2.2 Lu Na Receiving Guests with Tea

Lu Yu cited *The Book of Zhongxing in the Jin Dynasty* in Chapter Ⅶ of *The Classic of Tea* that: "When Lu Na was on the post of Wuxing Satrap, Xie An, General of Guardian Forces, came to visit Lu Na. Lu Chu, Lu Na's nephew, wondered why he did not prepare anything to receive Xie An, but he dared not ask the reason and he privately prepared a meal for more than ten persons. When Xie An arrived, Lu Chu found that only fruits and tea were there, and so he had to put all he had prepared to receive Xie An. After the departure of Xie An, Lu Na punished Lu Chu for forty blows, saying: 'Though you cannot enhance my good deeds, how can you damage my honest and upright integrity?' " It is also recorded in Volume VI "Tea Savoring" in *Observations of Feng Yan*: "Lu Na had nothing but fruits and tea to receive Xie An." This is another story in the history of Huzhou about receiving guests with tea and works as a typical example of clean government. Lu Yu exploited this story of Lu Na using fruits and tea to receive Xie An to show his uncorrupted proposition performance in daily life. The original book, *The Book of Zhongxing in the Jin Dynasty*, has been lost.

Lu Na (?-395), Wuxing descendant, style name Zuyan, the son of Lu Wan, was of ambition and integrity at young age and was far above the general public. When Emperor Jianwen was in power (371-372), he appointed Lu Na as Satrap of Wuxing Prefecture. He refused to receive the salary when he came to the post of Satrap of Wuxing Prefecture. When he left his post, he had nothing but a quilt. All the things except his quilt were sealed and returned to the government. When Emperor Xiaowu of the Eastern Jin Dynasty was in power (373-396), he was moved to Taichang (a post in charge of ceremonies of worshiping ancestors) and was appointed as Minister of Official Personnel Affairs. Xie An (320-385), styled name Anshi, served successively as Wuxing Satrap, Head of Ministers, Supervising Minister, General of Cavalry and Chief Minister. In Taihe Period (366-371), he served as Wuxing Satrap and once dug a canal in the west of the town (now Hongqiao Town, Changxing County) and benefited people greatly. This canal was also called Xiegongtang. In the eighth year (383) of Taiyuan Period, he was the commander of the famous Feishui Battle of China. When he finally won the battle, he was promoted as Head of Military Affairs. Xie An's family had deep relation with Huzhou City in that there were five Wuxing satraps in three generations. In the 11th year (579) of Taijian Period of the Chen Dynasty of the Southern Dynasty, when Xie Yiwu, Xie An's descendant, was appointed as Magistrate of Changcheng (now Changxing County of Huzhou City), he moved Xie An's tomb from Meishan Mountain of Jiankang (now Nanjing City, Jiangsu Province) to Yagang in the southwest of Changxing. Later it was built into Xie Taifu Temple by the successors.

1.2.2.3 Monk Fayao's Longevity for Tea Drinking at Xiaoshan Temple

In Chapter VII of *The Classic of Tea*, Lu Yu cited the book *Sequel to Biographies of Eminent Monks* by Monk Daogai, which recorded Fayao's story. Fayao was a monk of the Song Dynasty of the Southern Dynasty. His secular family name was Yang and was a native of the east to the Yellow River, that is, the middle of Shanxi. In the middle of Yuanjia Period, he crossed the Yangtze River and came across Shen Taizhen who was at Xiaoshan Temple of Wukang. Then, he was over 70 years old and drank tea for each meal. In the middle of Yongming Period, an order came from the Emperor that Fayao was invited to the capital (now Nangjing City) when he was 79 years old. Daogai narrated a story of longevity of a famous monk for drinking tea when he preached Buddhism to the monks and Buddhist people. Fayao, Zen Master, lived in the Song Dynasty of the Southern Dynasty. His secular family name was Yang before he became a monk. He came across the Yangtze River in the middle of Yuanjia Period, that is between

424 and 453, when Liu Yilong, Emperor Wen of the Song Dynasty, was in power. Fayao was thought highly of by Shen Taizhen who invited him to host Xiaoshan Temple in Wukang. Fayao was keen on learning from childhood. He had travelled ten thousand *li* to practice and could not only understand all the Buddhist books but also read books of different countries. In 422, the 19th year of Yuanjia Period of the Song Dynasty of the Southern Dynasty, he began to give lectures on Buddhism. At that time, the roads to Xiaoshan Temple were crowded with people coming to attend Fayao's lectures on Buddhism. All these people came from Sanwu (referring to Wujun, Wuxing, and Kuaiji in *The Chorography of the Three Kingdoms* and *History of the Jin Dynasty*, or referring to Wujun, Wuxing and Danyang in *General Laws and Regulations of All Dynasties* and *Records of Yuanhe County*).

When Fayao wrote such Buddhist books as *Nirvana* (mainly about all living creatures having Buddhist nature), *The Lotus Sutra* (mainly about tolerance of all beings), *Full Version of Buddhist Sutra* (mainly about requirements, procedures and ceremonies for becoming a monk) and *Shenghuan Sutra* (mainly about all world being Dharma Realm), he was "nearly 70 years old" and was described as "the sun going to set" in "Rules of Astronomy" of the book by Liu An, litterateur and thinker in the West Han Dynasty. However, Fayao drank tea at every meal and was very healthy.

In 462, the sixth year of Daming Period of the Song Dynasty of the Southern Dynasty, Liu Jun, Emperor Xiaowu, gave an edict to Wuxing to invite Fayao to the capital (now Nanjing City) and Fayao was arranged to lodge at Xin'an Temple. As soon as Fayao arrived in the capital, he began to give lectures on Buddhism and Emperor Xiaowu himself attended the lectures with all ministers of the palace. At that time Fayao was already 79 years old. According to "Lu Yu and Xiaoshan Temple of Wukang" by Cai Quanbao, Xiaoshan Temple of Wukang, also called Cuifeng Temple, was built between the third year (282) of Taikang Period of the Western Jin Dynasty and the first year (420) of Yongchu Period of the Song Dynasty of the Southern Dynasty. It was destroyed in the Yuan Dynasty. The relics are still in Yangkou Village, Wukang Town of Deqing County, Huzhou City.

It has to be pointed that the allusion in *The Classic of Tea* by Lu Yu happened in the Song Dynasty of the Southern Dynasty, but the time was mistaken based on the examination of different versions of *The Classic of Tea*. It was about 180 years between Yongjia Period (307-312) of Sima Chi, Emperor Huai of the Western Jin Dynasty and Yongming Period (483-493) of Xiao Ze, Emperor Wu of the Qi Dynasty of the Southern Dynasty. According to these two periods, Fayao should be about 200 years old, which meant that Fayao had lived across the Song Dynasty of the Southern Dynasty. The conclusion is that Yongjia should be corrected as Yuanjia (424-453), which has been proved by many other scholars.

1. 2. 2. 4 Lines of Poem by Bao Zhao of the Song Dynasty of the Southern Dynasty

Bao Zhao, poet and essayist, was the only celebrity of the Southern and Northern Dynasties that was mentioned twice in *The Classic of Tea* by Lu Yu. Lu Yu first mentioned Bao Zhao in Chapter Ⅶ of *The Classic of Tea* that Bao Zhao's sister, Bao Linghui, wrote "Ode to the Fragrant Tea". Then Lu Yu mentioned Bao Zhao in his *Record of Zhushan Mountain* that Miaoxi Temple was the place where Bao Zhao composed a poem to see Assistant Minister Sheng and Division Commander Geng off. Miaoxi Temple was a very good place where there was Huangpu Bridge twenty steps away from the temple and there was Huangpu Pavilion fifty steps in the south of the bridge.

Bao Zhao (c. 414-466), styled name Mingyuan, was from the Song Dynasty of the Southern Dynasty. He was born in a poor family and he was very talented but had a low position. He once felt pity on himself saying that in the history of thousands of years there were countless talents without the opportunity to serve the country. In 439, the 16th year of Yuanjia Period, when he chanted poems to announce his ideal, he was recognized by Liu Yiqing, Duke of Linchuan, promoted as Assistant Minister of his kingdom. After the death of Liu Yiqing, he had to leave his post and stayed at home without doing anything.

In 447, the 24th year of Yuanjia Period, Bao Zhao presented "Ode to the Clear River" to the palace and was promoted again as Assistant Minister by Liu Jun, Duke of Shixing. Many years later Bao Zhao was appointed as Magistrate of Yong'an County. When Emperor Xiaowu was in power, Bao Zhao was removed from as Magistrate of Haiyu County (now the east of Changshu City of Suzhou). Later he was promoted as Scholar of the Imperial College, Secretary of the

Emperor, Magistrate of Moling County (now Jiangning District of Nanjing City of Jiangsu Province), and Magistrate of Yongjia County (now Wenzhou City of Zhejiang Province). Then, Bao Zhao went to seek refuge with Liu Zixu, Duke of Linhai, the seventh son of Emperor Xiaowu, and was appointed as Military Adviser, which was why Bao Zhao was called Military Adviser Bao. According to the relationship Bao Zhao had maintained with Liu Zixu, the author of "Literati Living in Wuxing Prefecture" included in the book of *Wuxing Prefecture and the Culture of Distinguished Families* holds that it was from the year 454, the first year of Xiaojian Period of Emperor Xiaowu, to the year 460, the fourth year of Daming Period of Emperor Xiaowu, that Bao Zhao was living in Wucheng (now Huzhou City), the seat of Wuxing Prefecture. Meanwhile he wrote poems such as "Seeing Division Commander Geng Off at Huangpu Pavilion of Wuxing" "Seeing Assistant Minister Sheng Off at Jianhou Pavilion" and "Looking Eastward at Zhenze (Taihu Lake) on Top of Lishan Mountain". Bao Zhao was fond of tea drinking and left us very famous lines of "with clear fountain flowing and trickling, lush sweet tea trees abound in the wilderness".

Chapter 2

Rapid Boom:

Huzhou Tea Culture in the Sui and Tang Dynasties

2.1 The Basic Situation of Huzhou Tea Culture

The history of the Sui Dynasty is very short. It only lasted for thirty-eight years (581-618), and there are rarely any records about tea. The unification of China in the Sui Dynasty and the building of the Grand Canal connecting the north and the south laid the foundation for the development of economy, culture and tea production of the Tang Dynasty.

The boom of economy and culture and the strengthening of the national force as well as the liking for tea in the Tang Dynasty directly promoted the development of tea culture and the production of tea. *Observations of Feng Yan* written by Feng Yan of the Tang Dynasty records: "In the middle of Kaiyuan Period of the Tang Dynasty, there was an exorcist who lived in Lingyan Temple of Taishan Mountain. He was working very hard to preach Zen and required his followers to learn it day and night. He advocated that his followers should drink tea instead of having supper. And so they carried tea with them wherever they went and boiled it to drink. Others followed their suit and it gradually developed into a custom. This custom spread gradually from the cities of Zou, Qi, Cang, Di to the capital. Most cities had tea shops and the passers-by from all walks of life could take it if he or she was willing to pay a fee. The tea from the Yangtze River basin and the Huaihe

River basin was transported by water or land. It piled up like mountains with its rich colors." Feng Yan adds: "The ancients also drank tea but they were not so indulgent in tea day and night as we do now." There are two points worthy of attention: Firstly, in the middle years of Kaiyuan Period of the Tang Dynasty, that is 723-732 AD, "it was a custom that tea was served day and night". Secondly, "the tea from the Yangtze River basin and the Huaihe River basin" means that tea was produced in the lower reaches of the Yangtze River which is now the southern part of Anhui, Jiangsu Provinces and the northern part of Zhejiang Province with Huzhou Prefecture as its center.

The Classic of Tea by Lu Yu written in Huzhou in the late years further confirmed that Wucheng (now Wuxing and Nanxun Districts), Changcheng (now Changxing), Wukang (now Deqing) and Anji County of Huzhou in the Tang Dynasty were so abundant with tea that "it was transported to other places by boats and other vehicles in succession". In the 1980s, Tea Association of Zhejiang Province published *Zhejiang Tea*. It is further documented that "in the Tang Dynasty, tea production in Zhejiang had a considerable scale, and with obvious nature of commodity. Places such as Huzhou and Pingshui Town of Shaoxing were famous provincial tea distribution markets. Huzhou and Fuliang of Jiangxi Province were famous national tea trade centers".

During the Tang Dynasty, Huzhou Zisun Tea, with the help of Lu Yu's appraising, raised its fame and had a considerable social status. According to *Notes about Building the Tea House of Yixing County in the Tang Dynasty*, "The tribute tea of Yixing is not an old story. Before that, when Imperial Censor Li Qiyun was the prefect there, a mountain monk offered him some good tea and Li used it to entertain the guests. The civilian Lu Yu thought that its scented fragrance, sweet and spicy taste was better than those produced in other places and thus recommended that it be used as tribute. Qiyun followed the advice and paid a tribute of ten thousand *liang* to the court for the first time." "Emperor Daizong of the Tang Dynasty ordered Qiyun to construct places of a few acres for tea trees and pay the tribute to the court every year, and so tea trees were planted from mountain to mountain since the fifth year (770) of Dali Period of the Tang Dynasty."

Due to the fact that Zisun Tea from Guzhu valleys was very good because of "its purple buds, new shoots like those of bamboos, revolting green leaves, intoxicating fragrance, and the delightfulness when sipped", it was favored by Li Yu, Emperor Daizong, and the imperial family of the Tang Dynasty. According to *Annals of Huzhou Prefecture*, *Annals of Changxing County*, from the fifth year of Dali Period to the sixteenth year of Zhenyuan Period (770-800), Changxing of Huzhou established a tribute tea house at the foot of Guzhu Mountain at the back of Hutouyan. At first the house was comparatively small with only 30 simple rooms. "Because the rooms were simple and not in good condition, in the seventeenth year (801) of Zhenyuan Period, Prefect Li Ci built a temple ... with the 30 rooms in the east gallery serving as the tribute tea house with stone pestles used for mashing steamed tea leaves at both sides. It had about 100 tea baking stoves and 1,000 craftsmen." Later, as the amount of tribute tea increased, the tribute tea house became larger and larger. "It took about 30,000 laborers a few months to produce Zisun Tea for the imperial court every year at Guzhu." The total tribute tea houses and tea-baking rooms amounted to 130-150 with more than 1,000 persons who processed tea and 30,000 people who picked tea. The scale of picking and processing tea was rare in the whole country. According to *Annals of Wuxing* in Jiatai Period: in the history's most prosperous Huichang Period (841-846), the tribute of Huzhou Zisun Tea reached up to 18,400 *jin*, which equaled 12,162.4 kg if discounted to a modern scale. The amount of the tribute tea is the largest in the whole country.

A more stunning fact of Huzhou Zisun Tribute Tea is "prefects of the two prefectures would come in person when the tribute tea houses were processing tea". History records showed that nine prefects of Huzhou in the Tang Dynasty went to Guzhu Mountain to supervise the production of the tribute tea, namely, Yan Zhenqing in the seventh year (772) of Dali Period, Yuan Gao in the second year (781) of Jianzhong Period, Yu Di in the seventh year (791) of Zhenyuan Period, Li Ci in the 16th (800) year of Zhenyuan Period, Cui Yuanliang in the third year (823) of Changqing Period, Pei Chong in the ninth year (835) of Dahe Period, Yang Hangong in the third year (838) of Kaicheng Period, Zhang Wengui in the first year (841) of Huichang Period, Du Mu in the fourth year (850) of Dazhong Period. The nine terms of office lasted nearly eighty years.

Every year, Huzhou and Changzhou would grow tea in their mountains respectively and had a banquet at Zhuomu Ridge of Changxing, Huzhou, which is the dividing line of the two prefectures. At the top of this ridge there is Jinghui Pavilion. The poet Bai Juyi of the Tang Dynasty wrote a poem "On Hearing That Jia of Changzhou and Cui of Huzhou Were Having a Tea Banquet at Jinghui Pavilion at Night":

> Far away I am hearing of the banquet at night,
> The beauties and songs are all around.
> Though two prefectures are bordered under the disk,
> A spring gathering is going in front of the same lamp.

In order to supervise the manufacture of the tribute tea, every prefect of Huzhou had to go to tea-growing mountains before Spring Equinox and stayed there until Grain Rain when the baking of the tribute tea was finished.

The imperial family ordered that the first tribute tea should be on its way of transportation ten days before Clear and Bright and reach the capital of Chang'an within ten days. This was called "urgent tea" which was mainly used for worshipping the imperial ancestors in the temple. The rest of the tribute tea was required to be transported to the capital before the end of April. Zhang Wengui, one of the prefects of Huzhou, once wrote a poem depicting the grandeur when Huzhou tribute tea was sent to the capital.

In the ninth year (835) of Dahe Period, Pei Chong, who was the prefect of Huzhou at that time, was dismissed from his office for his mismanagement of manufacturing the tribute tea. From then on, all the prefects of Huzhou spared no effort and care when they were supervising the production of the tribute tea which the emperors liked.

In addition, Yuan Gao and Du Mu, prefects of Huzhou, both wrote poems of tea mountain.

Particularly worthy of praise is Yuan Gao, who was gentle and generous. People admired him because he had the courage of advising frankly. Seeing the toiling of the civilians who ceased their normal work of agriculture and sericulture to produce the tribute tea day and night when he was supervising the production of it, he wrote a "Poem of Tea Mountain" after the completion of his task that year and sent it together with the tribute tea to the imperial court.

2.2 Lu Yu's Main Activities in Huzhou

Lu Yu (733- c. 804), another name Ji, styled name Hongjian, also self-claimed to be an Old Man of Mulberry and Ramie, also Donggangzi, is a scholar in the Tang Dynasty. He is the first person who put forward systematically the methods of picking, making, boiling and drinking of tea. He was an ancient expert of agriculture in China. He wrote the first classic work about tea—*The Classic of Tea* and thus is worshipped as "the Tea Sage".

Born in the 21st year (733) of Kaiyuan Period of the Tang Dynasty in Jingling of Fuzhou (now Tianmen City, Hubei Province), Lu Yu was a deserted boy adopted by Zen master Zhiji of Longgai Temple. Through divinating by *The Book of Changes*, Zhiji got the words "*hongjian yu lu*—a giant wild goose lands slowly on the land", "its feather (*yu*) can be used as a ritual gift", and thus he took "Lu" as the boy's family name, "Yu" the given name, and "Hongjian" the styled name. Lu Yu spent his childhood in the courtyard of old temples. He used to boil tea for Father Ji and thus became very interested in tea. Later he joined a circus acrobat and worked as clowns in plays. He was appreciated by Satrap Li Qiwu of Jingling and was given some books and taught to do research work. Edified by the poet Cui Guofu who was Sima of Jingling in the Tang Dynasty, he plunged himself into poetry writing. Hence, when Lu Yu grew up, he was not only an expert at poetry and essays, but also a specialist in picking, making, boiling and drinking tea. He often gathered with his friends such as poetess Li Jilan and Monk Jiaoran to express their feelings and ambitions by drinking tea and chanting poems.

2.2.1 Seclusion in the Tiaoxi River of Huzhou

"The mountain is very high; it seems to be in the sky." More than 1,200 years ago, Huzhou was one of the most powerful and prosperous cities in the south of China and it extended along rivers and stopped at mountains. To the west there is majestic Tianmu Mountain; to the north lies Taihu Lake, a vast expanse of misty, rolling waters. In the 14th year (755) of Tianbao Period of Emperor Xuanzong of the Tang Dynasty, An Lushan revolted in Fanyang, and in the early years of Zhide Period (756-758) of Emperor Suzong of the Tang Dynasty, Lu Yu crossed the Yangtze River to the south in order to avoid the disorder. After a long journey, he came to Huzhou at the prime age of 24 and visited Jiaoran—a monk poet whom he had heard so much about in Miaoxi Temple in the southwestern suburb of Huzhou. Jiaoran (730-799), whose secular family name is Xie and given name is Qingzhou, was born in Changcheng (now Changxing, Huzhou). He was one of the tenth generation descendants of Xie Lingyun who created the school of landscape poetry in the Southern Dynasty. At that time Jiaoran was the presider of Miaoxi Temple. Jiaoran found that Lu Yu was elegant in conversation, proficient in classics and history, well informed of miscellaneous knowledge, and fond of poetry and tea. They became cross-age friends because they were like-minded. "The Autobiography of Lu Lu" in Vol. 793 of *Finest Blossoms in the Garden of Literature* writes: "In the early years of Zhide Period, I followed the northerners and crossed the Yangtze River to the south and made friends of different generations with Monk Jiaoran of Wuxing." *Collections of Wuxing Anecdotes* also records: "Jiaoran is a talented son of the Xie family of Huzhou. He is often associated with such persons as Yan Zhenqing, Yu Di and so on. When Lu Yu came, they chatted day and night, despising writing for fame. On the Double Nine Festival of that year, Lu Yu and Jiaoran made a poem 'Drinking Tea with Lu Yu on September the 9th': 'On the ninth of September in the temple, yellow is the chrysanthemum on the east fence. Secular people are all fond of wine, but who knows the aroma of tea?'"

At the beginning years (760-761) of Shangyuan Period, Lu Yu still "lived in seclusion in Zhushan Mountain of Wucheng" (refer to Qianlong's edition of *Annals of Huzhou Prefecture*).

During his absence from Miaoxi Temple in Zhushan Mountain, Lu Yu often went out to search for temples and tea-producing districts, and Jiaoran would always look at the moon when he was missing his friend. Once when Lu Yu came back, Jiaoran was so happy that he made the poem "Waiting under the Mountain Moon": "Night to night I miss my old friend, so I often wait for him under the mountain moon. Tonight my old friend came; Does the mountain moon know where he is? "

Huangqian Ridge of Shangbai Village in Deqing County was once named Hongjian Peak. Legend goes that Lu Yu once lived there when he investigated the tea district of Wukang. Later descendants named both the village and the ridge "Hongjian" in memory of him. And so now there is still a Hongjian Village. Hongjian Peak is situated face to face with the famous Mogan Mountain.

During the period of Dali (776-779), Lu Yu lived in Qingtang Village which is outside Yingxi Door (also Qingtang Door) of Huzhou. Qingtang Village is located 1.5 kilometers northwest of Huzhou. Sun Xiu, Emperor Jing of the Wu Kingdom of the Three Kingdoms Period, built copper mills there which extended westward tens of *li* approximately from Yingxi Door to Changxing, and because of this the village was named Qingtang Village. It is on the south slope of Qishan Mountain and near the side slope of Fenghuang Mountain. To the west of it is the area where the East Tiaoxi River and the West Tiaoxi River intersects. Ye Mengde, a writer of the Southern Song Dynasty, wrote a poem: "The mountain shapes like a crown; It looks similar from every side. Near the gate of Wucheng County; The Tiaoxi River and the Zhaxi River share the same water source." As early as the first year (627) of Zhenguan Period of the Tang Dynasty, this area had been producing mulberry and hemp, and "silk can be produced anywhere, but Huzhou produces the best" "in Huzhou, every family cultivates ramie and makes it into thread which is mostly woven into cloth". Therefore, "Biography of Lu Yu" from "Hermit Part" in *New Book of Tang's History* records that Lu Yu "secluded in the Tiaoxi River catchment, represented himself as an Old Man of Mulberry and Ramie".

After Lu Yu settled in Qingtang Retreat, Li E, Jiaoran and Quan Deyu went there as guests. They chanted and drank tea together, and even "sat up at night and chatted about old times". Jiaoran wrote a poem "Gather with Li E the Attendant in the Palace at Recluse Lu Yu's New Home":

> Your house is in a simple style, even a thousand houses are difficult to match yours,
> But the words you utter are novel and new.
> Moving home often makes one feel isolated,
> But changing dwelling places also makes you stronger.
> You can fish and plant bamboo;
> You can cultivate vegetables all by yourself.
> You get brothers by martial arts;
> And make friends by chanting poems.
> There are passers-by under the thick willow trees;
> Fragrance of the deep paths reminds me of Zen monks.
> It's not that you are hidden by the east wall,
> But the guests have never been here before.

Li E had served as an attendant in the palace. From the 8th to the 11th years (773-776) of Dali Period, he served as Deputy General of Militia in Huzhou. Jiaoran also wrote a poem named "Not Finding Lu Hongjian at Home":

> I found you, moved beyond the city,
> A wide path led me, among mulberry and hemp,
> To a newly-set hedge of chrysanthemums,
> Not yet blooming although autumn has come.
> I knocked, but no answer, not even a dog.
> I went to ask your western neighbor.
> He told me that you climbed mountains every day,
> Never returned until the sun set.

This is a wonderful ode to the Tea Sage. It makes the person extrapolate and have an immersed sense. According to *Complete Poetry of the Tang Dynasty*, Jiaoran wrote two poems in Lu Yu's residence. One poem is "Spring Evening

Gathering at Lu Yu's Home":

I can't wait until morning to enjoy the beauty of flowers,
Heartless men visits affected ones.
Is there no morning in Xilin?
It is because I forget our friendship and springtime.

Another poem is "Delighted in Yixing Prefect Quan Deyu's Arrival from Junshan Mountain and Accompanying Him to Lu Yu's Qingtang Retreat":

It is impossible for you to resign from an official position,
So it is better for you to cancel yourself temporarily as a hermit like Junyang.
I can see the plaque on the door of the county government building,
As well as the newly painted white powder on the wall.
I had come to drop on you,
But I was attracted by the beautiful bamboo.
The house is inlaid in the white clouds,
Open the door and you can have a panoramic view of the mountains.
It is the happiest thing to be carefree,
And to fish in the stream near the fence.

Prefect Quan, who is mentioned in the title of Jiaoran's poem, is Prefect Quan Deyu of Yixing that is adjacent to Huzhou. Quan Deyu (761 or 759-818), style name Zaizhi, was a native of Lueyang of Tianshui (now northwest of Qin'an, Gansu Province). He was trusted by Emperor Dezong because he was good at writing. He served as the secretary of the emperor for a number of years, and later took the position of the Minister of the Board of Rites. He wrote a lot of works and left *The Collection of Quan Zaizhi* to the world. Quan took up the post of Yixing Prefect when he was only 30 years old, and at that time Lu Yu was more than 50 years old. Due to the spread of *The Classic of Tea*, Lu Yu enjoyed a high reputation at that time. It is natural that a young local governor condescended to visit Lu Yu. Accompanied by Jiaoran, Quan Deyu made inquiries about Lu Yu; and he also wrote poems. One is "Chatting about Old Time with Old Friends at Night":

> We talked and laughed through tonight,
> Missing the days of the past when we played around the mountains and rivers.
> Under the moon we discussed the success or failure of the past.
> To the elderly, the career is like in the fall.
> No matter how the career is, let's be happy and pleased,
> No longer seek fame.
> Still wearing the original clothes,
> We are living in seclusion in the mountains.

After Lu Yu's death in Huzhou, Quan Deyu also wrote "Weep on Recluse Lu":

> We can't see each other any more from now on,
> The wooden door is opening in front of snow.
> Now I live in troubled times,
> You have hidden all the sorrow to the underworld.
> The light of the setting sun penetrates the lonely empty hall,
> Isolated and helpless city summons the guests to go back home.
> Looking for a true person is still a long way to go,
> And the letter of recruitment still does not come.

2.2.2 Investigating the Tea Area of Huzhou

When Lu Yu resided in Huzhou, he often went to mountain areas and temples to investigate tea shoots among bushes and mountain springs for boiling tea. According to *The Autobiography of Lu Yu*: "I often went to mountains and temples by boat, wearing a short brown shirt and pants, with a scarf over my shoulder, a pair of cane shoes under my feet. Mostly I would walk alone in the wild, chanting Buddhist scriptures and ancient poems, tapping trees with my walking stick or touching flowing water by hands, lingering around from day to night. I would not go back home until it was dark and I was tired out."

It is recorded that Lu Yu had been to many mountains of Huzhou: Guzhu Mountain, Fengting Mountain, Xiyan Mountain (Xuanjiao Ridge), Zhuomu Ridge, Qingxian Mountain, Shimen Mountain, Huangqian Ridge, Xiaoshan Mountain, mountainous areas in Anji County, Bianshan Mountain (Wenshan Mountain), Xisai Mountain, Xianshan Mountain, Zhushan Mountain, and Jin'gai Mountain.

In Chapter Ⅷ of *The Classic of Tea*, Lu Yu evaluates the quality of tea in the western Zhejiang Province and points out that "in the west of Zhejiang, Huzhou's tea is the best, Changzhou's is the second and the tea from Xuanzhou, Hangzhou, Muzhou and Xizhou is the third, and Runzhou's and Suzhou's tea is the fourth". Original notes: "In Huzhou, the tea grown in Guzhu Valley of Changcheng County is the same with the tea grown in Xiazhou and Guangzhou; the tea grown in the areas of Wushan Mountain, Tianmu Mountain, Baimao Mountain and Xuanjiao Ridge is the same with the tea grown in Xiangzhou, Jingnan and Yiyang; the tea grown in the areas of Fuyi Pavilion of Fengting Mountain, Feiyun Temple and Qushui Temple, Zhuomu Ridge is the same with that grown in Shouzhou and Changzhou; the tea grown in the valleys of Anji County and Wukang County is the same with that grown in Jinzhou and Liangzhou." *The Classic of Tea* refers to ten tea-producing areas of Huzhou.

Guzhu Mountain, which is located seventeen kilometers northwest of Changxing County, Huzhou, is 355 meters above sea level, and has an area of about two square kilometers. The mountain got that name because the younger brother of Fuchai (the King of Wu) ascended the east of Guzhu during the

Warring States Period. There is the boundary of Zhuoshe in the southern side of the mountain, and in the northern side, there is the boundary of Xuanjiu. It is a place where the cliffs rise steeply and waterfalls flow straight down. The tea grown there is exceptionally good. In Guzhu Mountain there is a tribute tea house in which there are Qingfeng Building, Zhenliu Pavilion, Xigong Pavilion, Jinsha Pavilion, Wanggui Pavilion and Papaya Hall. Beside the tribute tea house, there is Golden Sand Spring with excellent water. When Zhang Wengui, Prefect of Huzhou, spoke of three wonders of Wuxing, he made a poem: "When the grass below Qingfeng Building begins to grow, the tea leaves in the Moon Gorge also sprout." To the northwest of Guzhu Mountain is Fengting Mountain connecting Xiye Mountain, which was so named because the spring water flows north and then turns west. The high cliff made the water splash around everywhere just like sobbing. In the middle, there is Xuanjiao (suspending) Ridge which is named after the dropping ridge. Xuanjiao Ridge is 250 meters above sea level. It divides Yixing of Jiangsu Province and Huzhou according to the boundary of watershed, and it is also a military fort in ancient times. "In the Tang Dynasty, Wuxing and Piling Prefectures would grow tea in their separate mountains and the prefects of the two prefectures would hold a party at Jinhui Pavilion (also named Fangyan) each year. The middle of the ridge is the boundary of the two prefectures and there still remain the relics of the deserted pavilion." "It was the place where Sun Quan shot a tiger in the 23rd year (218) of Jian'an Period." Zhuomu Ridge, also called Twenty-Three Bends, is 400 meters above sea level, and the north of its watershed of this ridge belongs to Yixing, Jiangsu. "Zhuomu Ridge is adjacent to Xuanjiao Ridge ... Beneath the thick clouds, it has lots of woodpeckers. Therefore, the mountain was called Zhuomu (Woodpecker) Ridge."

Lu Yu almost went about all the valleys of Guzhu Mountain, Fengting Mountain, Xiye Mountain and Zhuomu Ridge. This area is a range of interchange mountains which connect Yixing, Jiangsu. To its east is Taihu Lake, and in the southwest, stands Longwang Mountain with an altitude of 1,578 meters. The region is blessed with plenty of rainfall and a mild climate. Its peaks and ridges with its bubbling stream, covered by green bamboo layer by layer, are all enveloped in the mist. All this together with its thick and rich soil makes it an ideal region for tea production. After the investigation of the tea area of Guzhu

Mountain, Lu Yu wrote *Tea Episodes of Guzhu Mountain*, and compiled it into Chapter Ⅷ: Producing Regions in *The Classic of Tea* on the basis of the quality of tea. According to *Dictionary of Places of Interest in China*, Lu Yu once grew tea trees at the foothill of Guzhu Mountain and wrote Chapter Ⅰ: Origins of Tea in *The Classic of Tea* in which he pointed out that "the tea buds in purple are better than the green one; the bamboo shoot-shaped tea is better than bud-shaped tea". This is the origin of the name of Zisun Tea (Purple Bamboo Shoot) of Guzhu Mountain.

Qingxian Ridge, formerly known as Qingxian Mountain, has an altitude of more than 200 meters. The east of the ridge belongs to Changxing, Huzhou; and the west of the ridge belongs to Guangde County of Anhui Province. Lu Yu inspected and tasted the tea there, and concluded that the tea from Qingxian Mountain and Zhuomu Ridge is the same as that from Shouzhou. Shimen Mountain, also called Yaoshi Mountain, is more than 400 meters above sea level. It is close to Guzhu Mountain and abounds with Purple Bamboo Shoot Tea. Jiaoran had been there with Lu Yu and wrote a poem "Walk in Guzhu—to Pei Fangzhou":

> Sparsely populated Yaoshi is rich in Purple Bamboo Shoot Tea;
> However, who really knows which one is purple bamboo shoot-shaped, which one is the green bud.

The poet Pi Rixiu of the late Tang Dynasty also made a journey to Yaoshi Mountain and composed a poem:

> To find Yaoshi Mountain,
> I entered deep into the flatlands by the mountain.
> In the mountain, the size of the tea garden is so big,
> No one can remember how many acres are there.
> The spring water is very clear;
> The peak is surrounded with the cloud.
> In early summer, it rains a lot;
> But in the misty rain, the whole valley is full of camellia.

Lu Yu together with Jiaoran also had been to Changcheng (Changxing County now). Changcheng, also called the City of King Wu, is bordered by Taihu Lake in the east, and Wushan Mountain in the south, and "with the Tiaoxi River as its drainage; legend has it that the King of Wu once saw his daughter off there". Jiaoran mentioned the place in his poem "Visiting Lu Yu the Recluse". Jiaoran and Lu Yu also lodged at Gratitude Temple which is 500 meters away in the northwest from the town. Jiaoran composed a poem "Visiting Gratitude Temple on Cold Food Day and Lodging at the Room of Xiegong":

> There is an ancient temple on the hill;
> A man has lived there for many years.
> The temple is in the deep mountain surrounded by many ancient pines.
> Buddhist dharma is silence,
> On Cold Food Day, it is sparsely populated and I feel lonely.
> If there are dark rooms everywhere,
> For whom the sun shines?

Gratitude Temple was built in the first year (560) of Tianjia Period of Emperor Wen of the Chen Dynasty of the Southern and Northern Dynasties and was renamed Daxiong Temple in the second year (1065) of Zhiping Period of the Song Dynasty. It was moved to the southeast of the county about two hundred paces in the second year (1369) of Hongwu Period in the Ming Dynasty, and then it was abandoned.

Lu Yu had also been to Huangqian Ridge and Xiaoshan Mountain in Wukang, Deqing County. Huangqian Ridge is also called Hongjian Peak, and at the foot of the mountain, there is Hongjian Village. It was said that Lu Yu once lived there when he investigated the tea area of Wukang. In Chapter Ⅶ: Records and Anecdotes of *The Classic of Tea*, Lu Yu mentioned "Xiaoshan Mountain of Wukang" and he also told a story that a Buddhist monk named Fayao would drink tea whenever he took his meal. Lu Yu had also been to Anji which is adjoining to Wukang to investigate tea. When talking the quality of tea in Chapter Ⅷ:

Producing Regions of *The Classic of Tea*, he pointed out that "the tea grown in the valleys in Anji and Wukang is similar to the tea grown in Jinzhou and Liangzhou".

Bianshan Mountain, covering an area of 70 square kilometers, locates about 9 kilometers northwest of Huzhou. A saying goes that "the 72 peaks of Bianshan Mountain are like a lotus flower in the sky". The main peak which is at an elevation of 521.6 meters is called Yunfeng Top. Lu Yu made a detailed investigation of Bianshan Mountain. In *Topography of Wuxing with Illustrations*, he illustrates that "the origin of the name of Bianshan Mountain is because there are some people whose surname is 'Bian' (卞) lived here". He also said that "bian" (弁) has a homophone (卞) and they are interchangeable. Wenshan (the Chinese meaning is warm) Mountain which is one peak in Bianshan Mountain got its name because there is a warm spring in it. Lu Yu mentioned Wenshan Mountain in Chapter Ⅶ: Records and Anecdotes of *The Classic of Tea*.

Xisai Mountain is located 10 kilometers west of the city of Huzhou. Zhang Zhihe once fished in the Tiaoxi River at the northern foot of Xisai Mountain. When Yan Zhenqing was on duty of Prefect of Huzhou in the seventh year (772) of Dali Period, Lu Yu used to drink and chant with Yan Zhenqing, Zhang Zhihe, Xu Shiheng, Li Chengju and so on in Xisai Mountain. As the host, Zhang Zhihe led the chanting of poems and wrote five poems of "Ode to a Fisherman". One of them said:

In front of Xisai Mountain, you can see egrets are soaring freely in the sky;

Plump fish are swimming cheerfully in the water and peach blossoms are bright and full.

A man wearing blue bamboo hat and green cloak is fishing contently.

He is so fascinated by the beautiful spring scenery that he forgets to go home when it rains.

At that time everyone wrote five poems, totally twenty-five. Now, except Zhang Zhihe's five poems, the rest has no longer existed.

Zhang Zhihe (about 730-810), whose style name was Zitong and old name was Guiling, native of Jinhua of Wuzhou (now belongs to Zhejiang), was a poet of the Tang Dynasty. In Suzong Period, Zhang, once served as a Waiting Hanlin (a post in the Imperial Academy), but later he was dismissed from his office because of wrongdoing. Since then, he never took any official position and lived in seclusion in Xisai Mountain in Huzhou, and called himself the Fisherman of the Yanbo River. He was narcissistic and never behaved like a time-server. For a long time he often wandered between the green hills and blue waters and led a poor and easeful life. Once Lu Yu asked Zhang Zhihe whom he often contacted with when Lu Yu, Pei Xiu and Zhang Zhihe sat side by side and talked intimately. Zhang answered, "I live with air and take the moon as my lamp, and never say goodbye to brothers all over the world. How can I contact my friends?" This kind of communication and friendship was specifically recorded by Yan Zhenqing when he wrote "The Inscription to Zhang Zhihe—the Wanderer".

Xianshan Mountain is 2.2 kilometers south of Huzhou. "The first thing you see in the mountain is called Xian. The mountain is so named because as soon as you walk out of Dingan Door you will see it." Xianshan Mountain, which stands face to face with Fuyu Mountain, is peculiarly rugged and towering with the foot of it bathed in the Jade Lake. On the mountain, there is a goblet-shaped stone which can hold wine. In Kaiyuan Period of the Tang Dynasty, an official named Li Shizhi came to Xianshan Mountain and saw the cave. Therefore, he built an engraved slab and named the stone "Wazun" (engraved drinking cup). "Cangshi Wazun" is one of the eight scenic spots in Huzhou. In the eighth year (773) of Dali Period, Lu Yu, Yan Zhenqing and other 27 persons made a tour to Xianshan Mountain and wrote an antiphon poem of "Stone Goblet".

Zhushan Mountain is located 12.5 kilometers southwest of the city of Huzhou. There is Miaoxi Temple on the mountain which is the first place Lu Yu lived when he came to Huzhou. Originally there was Huangpu Bridge in front of the temple, and in the south of the bridge there was Huangpu Pavilion. In the east of the temple there was Zhaoyin Courtyard. In the west of the fore-room was a warm loft and in the southeast was a cliff. There was a fishing table near the cliff. Bita City was in the northwest of the temple. The temple is surrounded by lush trees and tall bamboos all round. With its elegant surroundings and beautiful scenery, the area is well-known for its three things—tea, bamboo shoots and osmanthus trees with red, green and purple flowers. In the seventh year (772) of Dali Period, Yan Zhenqing, Prefect of Huzhou, invited Lu Yu and other 18 scholars to edit *The Complete Collection of Rhymes*. At that time, "they discussed in the school of government in summer and went back to Zhushan Mountain in winter and finally they finished the compilation of the whole book in the next spring (773)". In October of Guichou (773), a pavilion was built in the southeast of the temple. According to Yan Zhenqing's "Inscription in Miaoxi Temple in Zhushan Mountain": "The name of this pavilion was called Sangui Pavilion (Three Gui Pavilion) because the building was built on Guihai day of Guimao month of Guichou year (October 21st, 773)." "Sangui Pavilion was built by Duke of Lu (Yan Zhenqing) for Lu Hongjian in Zhushan Mountain." (refer to *Places of Interest*) "Sangui Pavilion was built by Yan Zhenqing and named by Lu Yu the Recluse because it was built on *guimao* day of *guihai* month of *guichou* year." (*Gui* is a way of recording time in ancient China) (refer to *Annals of Wuxing* in Jiatai Period of the Southern Song Dynasty). "Inscription on Sangui Pavilion in Zhushan Mountain" by Yan Zhenqing says: "Why built Three Gui Pavilion on Zhushan Mountain? Actually it's for the sake of Lu Yu." Below the poem there is a short preface which says: "The pavilion was named by Lu Hongjian." Jiaoran poetized appointed lines in reply to the poem "Stepping on Sangui Pavilion of Miaoxi Temple by Yan and Lu". Below the poem there is still a short preface: "The pavilion was named by Lu." From the above historical descriptions, Yan Zhenqing, Prefect of Huzhou, built the pavilion in honor of those men of letters led by Lu Yu for their accomplishment of a vast project of compiling the 360 volumes of *The Complete Collection of Rhymes*. In the process of compiling these books, they held many assemblies, fetes, rafting on the Tiaoxi River and so on. Evidences can be found in many works of Jiaoran.

Jin'gai (Golden Cap) Mountain "got its name because it shapes like a golden cap". Later it was renamed Heshan Mountain because He Kai, Satrap of Wuxing, once studied Confucianism here. It is situated 9 kilometers south of the city of Huzhou, next to the East Tiaoxi River. It is 292.6 meters above sea level and has beautiful scenery. There are sweet springs, luxuriant trees, bamboos and tea bushes, splendid temples and mysterious fog all over the mountain. Lu Yu once went sightseeing there with his companions, and boiled tea with the water from

Cloud Spring. As Qianlong's Block-Printed Edition of *Annals of Huzhou Prefecture* goes: "Lu Yu wrote another volume named *Tractate on Water.*" He wrote: "Jin'gai was covered by clouds and fog in this time of the year. I went for a walk in Jin'gai in the morning with my companions. There were a lot of clouds and fog all over ... Later, the fog disappeared gradually and the sun rose with only Jin'gai covered in the fog for a much longer time." According to *Annals of Jin'gai Mountain*, Lu Yu wrote in *Tractate on Water*: "... The fog covered all of the mountains, and the terrain between those mountains was plain and humidified by it. After a while the sun came up and took away most of the fog, but the fog over Jin'gai Mountain remained for a long time. Nobody knew why. I thought the reason why there was a lot of fog was because there may be a good spring in it. So I went south, crossed some slopes and found out there was a spring flowing under a tree. I made some tea with the water of it, and discovered that the tea didn't change color. Natives said that the local people within 10 *li* all fetched water from this spring to boil silk, and the silk produced this way was bright and clear, and sold well. The spring irrigated about 100 *mu* of field at the foot of the mountain. It was called Cloud Spring." Cloud Spring also has other names such as White Cloud Spring and Milk Spring. It is at the north foot of Jin'gai Mountain.

2.2.3 Writing *The Classic of Tea* in Huzhou

It was more than forty years from the time Lu Yu arrived in Huzhou at the beginning of Zhide Period to his death in Huzhou at the end of Zhenyuan Period. During this period, the places he had been to or lived in are: Shaoxing, Yuhang and Huzhou of Zhejiang Province, Suzhou, Wuxi, Yixing, Danyang, Nanjing of Jiangsu Province, Shangrao, Fuzhou of Jiangxi Province and so on. He had travelled all over the vast regions of the middle and lower reaches of the Yangtze River. But one point that can be certain is that Lu Yu has spent the longest time of his life in Huzhou. His investigation of tea areas in Huzhou is the most detailed, and most of his works including poems, articles and books are written in Huzhou. According to records of *Annals of Huzhou Prefecture*, *Collections of Wuxing Anecdotes*, *Annals of Wuxing* in Jiatai Period, *Annals of Wucheng County*, *Annals of Changxing County*, *Annals of Jin'gai Mountain* and the work of "Inscription in Miaoxi Temple in Zhushan Mountain" "Xiting Notes by Wuxing Satrap Liu Yun of the Liang Dynasty of the Southern Dynasty" and "The Notes of Tablet of King Xiang" which were written by Yan Zhenqing, Lu Yu had written more than nineteen works in Huzhou. Namely, they are three volumes of *The Contract between Ruler and Subjects*, thirty volumes of *Origin of Family Surnames*, ten volumes of *Spectra of the Four Family Names in the South of the Yangtze River*, ten volumes of *The Figures of the Southern and Northern Dynasties*, three volumes of *Records of Wuxing Officials of All Ages*, one volume of *The Records of Prefects of Huzhou*, three volumes of *The Classic of Tea* and three volumes of *Oneiromancy*, several volumes of *Topography of Wuxing with Illustrations*, *Records of Guzhu Mountain*, *Records of Wuxing*, *Records of Zhushan Mountain*, ten volumes of *Foreboding Year*, ten volumes of *Poor God*, one volume of *Notes of Tea*, one volume of *Records of Musical Education Institution*, several volumes of *Tractate on Water*, *Discussion on Cursive Script between Monk Huaisu and Yan Zhenqing* and *Autobiography of Lu Yu* and so on. We can discern from the titles of the works that there are at least seven kinds of books that directly describe Huzhou.

Lu Yu had revised the scripts of *The Classic of Tea* once and again when he wrote it in Huzhou. In his autobiography, he said that he wrote three volumes of *The Classic of Tea*, on "the twenty-nineth day in September in Xinchou year of Shangyuan Period". It means that the first draft of *The Classic of Tea* was completed before the second year (761) of Shangyuan Period. The first revision was finished after the second year (764) of Guangde Period. In Chapter Ⅳ: Boiling Apparatus, he mentioned that "The wind furnace is cast from copper and iron, and it looks like an ancient tripod …'cast in the following year after Tang extinguished Hu' in one of the furnace feet". It points out specifically that the tool (the wind furnace) for making tea was molded in the following year after the suppression of the rebellion of An Lushan, or the second year of Guangde Period. It shows that Lu Yu had revised *The Classic of Tea* three years later and added to it such contents as "cast in the following year after Tang extinguished Hu". The third edition of *The Classic of Tea* was completed after the eighth year (773) of Dali Period. According to Chen Shidao (1051-1101) of the Song Dynasty, there were four versions of *The Classic of Tea* with different content which he had read. Some of them were simple, while others were more detailed, especially "Chapter Ⅶ: Records and Anecdotes". So it is reasonable that the part of "Records and

Anecdotes" was modified most. Yan Zhenqing created the outlines of *The Complete Collection of Rhymes* when he was in Pingyuan. When he became Prefect of Huzhou, he invited Lu Yu and other 18 literati in the south of China to "modify the old chapters, search for evidences widely, and accomplish three hundred and sixty volumes" "it was completed in the spring of *guichou* (the eighth year of Dali Period)".

Lu Yu obtained sustenance from the process of writing and compiling of *The Complete Collection of Rhymes* and added it to the seventh chapter of *The Classic of Tea*. Actually, "Chapter Ⅶ: Records and Anecdotes" recorded celebrities who were fond of drinking tea and allusions about tea in the past dynasties by "searching for evidences in books of sea". In *The Classic of Tea*, Lu Yu mentioned such anecdotes and celebrities as Sun Hao, the yielded king of the Wu Kingdom, Lu Na of Wuxing and his brother's son Lu Chu the civil official in Kuaiji, Xie Anshi the general, Monk Fayao of Xiaoshan Temple in Wukang, Shan Qianzhi of Henei and so on.

In addition to *The Classic of Tea* and other works, there are plenty of antiphon poems which were written by Lu Yu and his friends. They are "Antiphon upon Boating with Lu Youping in Autumn" and "A Double Antiphon" which were jointly written by Huzhou Prefect Lu Youping, Jiaoran, Pan Shu and Lu Yu. "Antiphon of Ascending Xianshan Mountain and Appreciating the Stone Goblet of Left Prime Minister Li Shizhi" "Antiphon of Wind with Gen Wei at Water Pavilion" "Antiphon of Listening to Cicadas in Youxi Station" "Three-character Antiphon of Enjoying the Moon at the South House with Huangfu Ceng" "Seven-character Double Antiphon" "Water Hall Antiphon to Deputy Magistrate Pan for Fun" "Antiphon of Bamboo Mountain Inscription about the Book Hall of the Pan Family" and "Antiphon of the Drunk Words" which were jointly written by Huzhou Prefect Yan Zhenqing, Li E, Jiaoran, Gen Wei, Huangfu Ceng and Lu Yu. "Antiphon about Long Distance" "Antiphon about Secret Longing" "Antiphon about Pleasure" and "Antiphon of Restricted Title" which were jointly written by Jiaoran, Yan Bojun and Lu Yu. "Antiphon in Leisure Time Presented to Hermit Lu San" which was jointly written by Gen Wei and Lu Yu. The above fifteen poems which Lu Yu participated can all be found in the ninth volume of *The Complete Collection of Tang Poems* and *Additional Works of Complete Collection of Tang Poems.*

In the Tang Dynasty, drinking tea, reciting poems or chanting antiphon were very popular and elegant activities among literati. It opened a new path for the Bailiang-style (Emperor Wu of the Han Dynasty in China built a stage named Bailiang and ordered each of his subjects to chant a line of a poem. Because of it, the Bailiang-style became famous).

2.3 The Main Content and Significance of *The Classic of Tea* by Lu Yu

The Classic of Tea is Lu Yu's achievement to which he devoted all his energy and wisdom. It is not only a natural science monograph of tea, but also a treasury of tea culture. The book is divided into three volumes and ten chapters, with about more than seven thousand words. The contents of the first volume are: 1) Origins of Tea, 2) Picking and Baking Tools, 3) Picking and Baking Methods; the contents of the second volume are: 4) Boiling Apparatus; the contents of the third volume are: 5) Tea Boiling, 6) Tea Savoring, 7) Records and Anecdotes, 8) Producing Regions, 9) Dispensable Tools, 10) Scroll Transcription.

The first chapter of *The Classic of Tea* talks about the origins of tea. Firstly, it aptly points out that tea originated from the south of China where there are giant tea trees with the height of dozens of feet and the width of two-men hug. Secondly, the book traces the historical origins of Chinese tea through the investigation of the various names of tea and its wording before the Tang Dynasty. Thirdly, this book gives an account of the relationships between the quality of tea and the growth of tea plants, the methods of planting, the soil, the environment in which the tea grows as well as the efficacy of tea and its negative effects, etc.

The second chapter of *The Classic of Tea* talks about the tools of tea picking and baking. It describes the styles, specifications, materials, methods to use, and precautions of the 16 kinds of tools required in the process of tea picking, processing and storing. It also makes a brief introduction of the seven processing crafts of tea in the process of picking, steaming, mashing, molding, baking, stringing and sealing.

The third chapter of *The Classic of Tea* is about methods of tea picking and baking. It focuses on expounding the suitable time of picking, weather conditions and the shape characteristics of selected fresh leaves. Next, it describes in detail the relationship between the appearance characteristics of eight kinds of finished teas and the quality of tea. Lu Yu believed that the best way to distinguish whether the tea is good or not lies in the practical experience of the farmers, that is "whether the tea has a high quality or not depends on the rules in heart".

The fourth chapter of *The Classic of Tea* is about tea bioling apparatus. It introduces in great detail the shapes, specifications, functions and usage of 28 kinds of utensils used for boiling and drinking tea. These utensils are divided into eight types according to their specific usage. It also tells the correct methods and principles of boiling and drinking tea. Moreover, it evaluates how the color and quality of porcelain tea sets in the Tang Dynasty affect the color of tea. Lu Yu boasted that the tea he boiled was as good as the soup that was made by Yi Zhi, the chancellor of the Shang Dynasty.

The fifth chapter of *The Classic of Tea* is about tea boiling. It emphatically discusses the skills such as baking tea, using firewood and water, boiling and pouring tea in the process of boiling tea. The discussion of using water is particularly penetrating and insightful. It emphasizes that good water is needed to boil good tea. It claims that "mountain spring water is the best, river water the second, well water the worst". There are some scientific points in his claim because mountain spring water which contains many minerals and grass fragrance is beneficial for human bodies and may have some influences on the color, aroma and taste of boiled tea to a certain degree. As for boiling, Lu Yu stressed on "the three boiling". The first boiling: water is boiled with bubbles like the eyes of fish in a slight sound. The second boiling: constant flowing bubbles are like spring springs. The third boiling: the water is like the flowing wave and raging tide. Wen Tingyun, litterateur in the Tang Dynasty, considered that "the method of the three boiling wouldn't be successful without the use of live fire" and only in this way can the tea retain its best essence, color and taste. Lu Yu said, the taste of real good tea must be bitter at first sip and later turns sweet in the throat.

The sixth chapter of *The Classic of Tea* talks about guidelines for drinking tea.

Firstly, it elaborates that tea drinking originated from the legendary Shennong, Emperor of the Five Grains. And there are records of tea in the Western Zhou Dynasty. It is not until the Tang Dynasty that tea flourished. Secondly, it talks about the category of tea, such as coarse tea, loose tea, powder tea and cake tea as well as the nine steps that are comparatively difficult to handle in the process of boiling and drinking tea. And finally, the relationship between the number of times of tea drinking and the number of tea drinkers is elaborated. What is interesting is that Lu Yu did not foresee the fashion that the cake tea, once regarded as a top grade in the Tang Dynasty, is now debased as "*ancha*" (inferior tea) while brewing tea with hot water directly has become the most popular and fashionable way of drinking tea nowadays.

The seventh chapter are records and anecdotes of tea. It provides an overview of the relevant historical events, allusions and poems before the Tang Dynasty. In this chapter, Lu Yu quoted from the historical records, biographies, notes, edicts, letters, medical books, exegetic books, geological books and mystery novels in the Western Zhou Dynasty, the Spring and Autumn Period and the Warring States Period, the Qin and Han Dynasties, the Three Kingdoms Period, the Western and Eastern Jin Dynasties and the Southern and Northern Dynasties, along with 47 records of tea events in 45 ancient books. Among the enumeration of 43 people who are related with tea events, there are many celebrities such as Ji Dan, Duke of Zhou, a politician in the Western Zhou Dynasty; Yan Ying, a politician in the Spring and Autumn Period; Sima Xiangru, a litterateur, and Yang Xiong, a philosopher in the Western Han Dynasty; Wang Su, a learner of Confucian scriptures in the Wei Kingdom of the Three Kingdoms Period; Zhang Zai, a man of letters, Zuo Si, Huan Wen and Sun Chu, essayists; Guo Pu, an etymologist in the Jin Dynasty; Bao Linghui, a poetess, Shan Qianzhi, a historian and Tao Hongjing, a pharmacologist in the Song Dynasty of the Southern Dynasty and so on. Some of the ancient books about tea that Lu Yu cited still remain, but there are some missing, so it's very precious that the cultural and historical heritage of tea can be preserved by *The Classic of Tea*.

The eighth chapter is about tea-producing regions. It mainly talks about the regional distributions of tea production in China. In the first year (627) of Zhenguan Period of the Tang Dynasty, the whole country is geographically divided into ten administrative Dao according to convenience of the location of mountains and rivers. In the twenty-first year (733) of Kaiyuan Period, the number of Dao increased to fifteen. Lu Yu lists 43 states and prefectures of the 8 Dao that are the main tea production areas in the south of China. According to the present administrative divisions, these areas include 15 provinces, cities or autonomous regions, namely Yunnan, Guizhou, Sichuan, Chongqing, Zhejiang, Jiangsu, Hunan, Hubei, Jiangxi, Anhui and etc. When Lu Yu talks about the five Dao of Shannan, Huainan, Zhexi, Jiannan and Zhedong, he also lists the names of the prefectures, counties and places where tea is yielded. What's more, he also divides the quality of tea into four levels: the upper, middle, lower and bottom.

The ninth chapter talks about dispensable tools. Lu Yu thinks that some tea-processing procedures or utensils of making and boiling tea can be omitted if tea is made or boiled in the Cold Food Festival when fire is forbidden or in an open temple garden, or on forest stones, or in a rock cave by mountain streams or in any other special environment. But in formal tea feasts, 24 kinds of utensils used for boiling and drinking are all required; otherwise the goodness of drinking tea will all be lost.

The tenth chapter talks about the scroll transcription of the book. It refers to the silk transcriptions hung on the wall. The content of *The Classic of Tea* is written on the plain woven fabric which is made up of the mulberry silk that is bounded in Huzhou in the Tang Dynasty. With its straight and smooth texture, this kind of silk is specifically used for calligraphy, painting and decoration. The transcriptions are placed or exhibited in the corners or on the wall so that people can remember and recite it whenever they enjoy a cup of tea.

The Classic of Tea written by Lu Yu has a profound cultural connotation and epoch-making significance in the history of tea culture in China and the world.

Firstly, *The Classic of Tea* infiltrates knowledge of *The Book of Changes* which is a classic book in the history of China.

Lu Yu carved *Xun*, *Li*, *Kan* ("巽""离""坎"), three of the eight trigrams in *The Book of Changes*, on the wind furnace which was one of the tools for making and boiling tea, combining the working process of blowing, ignition and boiling water with *The Book of Changes* to sublimate the spirit of tea. The fifty-seventh *Gua* in *The Book of Changes* is *Xun*. *Xun* is wind. With humbleness and obedience as its essence and lenience as its function, it has the meaning of concordance and mutual comprehension. The thirtieth *Gua* in *The Book of Changes* is *Li*. *Li* is fire and fire means brightness. That is the sun shines on the earth, and all things grow with it. If you take the doctrine of *Li* you will be just, selfless, honest and modest. The twenty-ninth *Gua* is *Kan*. *Kan* is water and water is the source of life. Everything comes from water and can be washed in the water. As far as personality is concerned, *Kan* is the mind and spirit of bravery. *The Book of Changes* uses eight trigrams (eight natural phenomena: sky, ground, thunder, wind, water, fire, mountain and marsh) to speculate on the changes of nature and society. It thinks that the interaction between *Yin* and *Yang* is the root of all things and proposes the simple dialectic viewpoint of "the strong and the weak (lines) displaces each other and produces change and transformation". Lu Yu fuses *The Classic of Tea* with *The Book of Changes* in order to teach people to learn to be a gentleman.

Secondly, *The Classic of Tea* expounds tea drinking under the theory of mutual promotion and restraint between the five elements and puts forward the idea of "balancing five elements to cure all diseases". It embodies the thought of natural harmony, healthy and upward mood.

Ancient Chinese thinkers believed that nature is composed of the five basic elements—gold, wood, water, fire, earth. They constructed the theory of "the five elements originate from each other and counteract with each other" to explain the unity of the origin and diversity of all things in the world. To originate from each other means to promote mutually. For example, "wood generates fire, fire generates earth, earth generates gold, gold generates water, and water generates wood". Mutual restrain means to restrict each other. For example, "water beats fire, fire beats gold, gold beats wood, wood beats earth, and earth beats water". These are views of simple materialism and spontaneous dialectics. The five elements property in ancient traditional Chinese medicine centers on the five internal organs of the body. So Lu Yu's view of "balancing five elements to cure all diseases" shows that regular tea-drinking can balance the five elements, reconcile the internal organs, and cure all diseases so that good health and longevity is reached. Lu Yu also quotes a story from *Sequal to Biographies of Eminent Monks* to prove his idea. In the Song Dynasty of the Southern Dynasty, there was a monk called Fayao who lived in Xiaoshan Temple of Wukang, Wuxing. He was healthy because he used to drink tea when he was eating. Fayao was 79 years old when he was summoned to the capital by Xiao Ze, Emperor Wu of the Qi Dynasty of the Southern Dynasty then.

Thirdly, *The Classic of Tea* thinks that drinking tea is the same as "being exact and thrifty". It claims that frugality cultivates virtue and good virtue promotes one's merits and that one should behave properly to improve one's character, be simple, quiet and inactive. In *The Classic of Tea* Lu Yu listed allusions that had happened in the history of Huzhou. The book mentioned that Sun Hao, Duke of Wucheng, also one of the grandson of Sun Quan in the Three Kingdoms Period, drank tea instead of wine and Lu Na, Satrap of Wuxing in the Eastern Jin Dynasty used tea and fruit to entertain Xie An, the general of guardian forces. Replacing wine with tea and entertaining guests with tea and fruit was regarded as elegant customs and habits of honesty and incorruptibility.

And finally, *The Classic of Tea* is also a scientific work about tea. It not only involves biology, ecology, soil science and cultivation, but also discusses the crafts of tea making, tea brewing and tea drinking. Lu Yu said, "trees are like *gualu*; leaves are like camp jasmines; flowers are like white roses; seeds are like palms; pedicels are like lilacs; roots are like walnuts." The botany properties of the tea trees are vividly depicted by these comparisons. Lu Yu also said that "The best tea grows in the fully weathered rocks. The second best tea grows in gravel soil. The worst tea grows in loess." "As for tea shoots that grow in sunny slopes or under the shade of forest, the purple tea was better than the green one, the bamboo shoots shaped are better than the buds shaped, and rounded tea leaves are better than extended ones." It indicates that the quality of tea is related to the soil, location, illumination, temperature and humidity. If the required skills are not properly used in the cultivation of tea trees, tea trees will "produce no tea or the leaves do not flourish even if the tea trees are planted". The treatise fully reflects Lu Yu's rich knowledge of biology, pedology and cultivation of tea.

The Classic of Tea describes in great detail the names, nicknames, sizes and materials of the 16 appliances that are used in the process of tea production according to the sequence of using. It also introduces in a complete and perfect way the 24 kinds of tea sets that are used in boiling and drinking tea. The tea parties that were held in the tribute tea house located in Guzhu Mountain of Changxing, Huzhou and in Jinghui Pavilion where the prefects of the two prefectures gathered provided a good opportunity for Lu Yu to observe and study tea painstakingly.

We can see from the 24 kinds of tea sets that Lu Yu paid great attention to the art of brewing and drinking tea and thought that "everything from nature has its own feature". The best annotation to these ideas is *Song of Tea* which was written by Jiaoran who was Lu Yu's cross-generation friend and the presider of Miaoxi Temple. "The first sip drives away sleepiness and makes you refreshed and relaxed. The second sip clears your mind and you feel like rain sprinkle on dust. The third sip gives you the truth and there is no need to solve troubles painstakingly." "Who knows the truth from drinking tea that is so comprehensive and profound? Only Danqiu can understand the truth." "Danqiu Lu Yu looks down upon precious treasure and delicious food; he becomes an immortal through picking and drinking tea." Jiaoran evolved tea drinking in *The Classic of Tea* and the fairy tales of the immortal Danqiuzi into poems. And he sublimated the lite character of tea and the functions of relieving worries, achieving harmony and preserving nature of drinking tea to serve for health, for delight, and for character and thus to live as an immortal. The delicacy of tea, the nature of its taste, the clarity of tea, the exquisite tea sets and the thoughts which are well-mannered, virtuous, frugal, truth pursuing, subtle, moderate and neutral are the true meaning of Chinese tea ceremony and the quintessence in Lu Yu's *The Classic of Tea*.

Since the advent of *The Classic of Tea* in the middle of the Tang Dynasty, "tea has become so popular that everyone of high social rank drinks it". And now, tea has become one of the world's top three popular drinks. *The Classic of Tea* is a hot topic for tea scientists and scholars as well as tea lovers and is pursued and maintained by them in successive dynasties. It is imprinted and published officially or privately in so big a scale that it is a wonder in the history of tea in China that there are so many editions of it. According to investigations on the editions of *The Classic of Tea*, there are *Nation-Wide Collection of Books*, *New Collection of the Tang Poems*, *Notes of Studies*, *Book of Solving Problems*, *Book of Everything*, *Annals of Song* in the Song Dynasty; *Book of All Notes from the Han to Song and Yuan Dynasties* in the Yuan Dynasty; *Series of the Tang and Song Dynasties*, *Book of a Hundred Celebrities*, *Series of Epistemology*, *Miscellaneous Records of Shanku*, *Novels of Five Dynasties*, Sangzhu Cottage Version, Zheng Cong Version, *Unofficial History by Wang Qi in the Ming Dynasty*, *Edition by Cheng Fusheng in Wanli Period of the Ming Dynasty*, Edition by Sun Dashou in Wanli Period of the Ming Dynasty; in the Qing Dynasty, there are *Tales of the Tang Dynasty*, *The Series of the Tang Dynasty*, Yihong House Version, *An Extended Edition of Plants with Illustrations* and a handwritten copy of the Qing Dynasty; a handwritten copy by Lu Xun, *Chinese Tea Ceremony by Huang Dunyan*, *The Collection of Ancient and Modern Books*, *Chinese Version of Everything about Tea*, *The Complete Works of Lu Yu* by Zhang Hongyong, *History and Allusions about Tea* by Zhu Xiaoming, *Selected Historical Documents of Chinese Tea* by Zhu Zizheng, *Interpretation of The Classic of Tea* by Cai Jiade, *Correction and Annotation of The Classic of Tea* by Zhou Jingming, *Notes about The Classic of Tea* by Deng Naipeng, *The Translation and Annotation about The Classic of Tea* by Fu Shuxun. Altogether there are about 41 versions. In 2000, in order to commemorate Lu Yu's 1200th anniversary, entrusted by Lu Yu Tea Culture Research Association of Huzhou, Ding Kexing took *Siku Quanshu* (*The Complete Works of Chinese Classics*) as the foundation, proofread and consulted various versions, and published *The Checking and Annotation about The Classic of Tea in Huzhou.*

The Classic of Tea was introduced to foreign countries in the fifteenth century. In Japan, there were Edo Version, "He" Version in Genroku Period, Baoli Version, Complementary Version in Tianbao Period, *Amplification of The Classic of Tea* in Yongan Period, Mikasa Study in Shouwa Period, *Annotation and Comments of The Classic of Tea* in Shouwa Period.

Since the beginning of the twentieth century, Lu Yu's *The Classic of Tea* is widespread in the world and has been translated into Japanese, Korean, English, French, Russian, German and other languages. There are 34 kinds of *The Classic of Tea* stored in Japanese National Congress Library, the Library of Congress of the United States, London University Library and some private libraries. *The Classic of Tea* was included in *Encyclopedia Britannica* in 1928. Dr. C. R. Harvard, a British tea scientist, says in *Manufacture of Tea* that "Brewing tea as a drink has a long history and the first authoritative work about tea—*The Classic of Tea* was written by Lu Yu of China". William Ukers, an American scholar, writes in *Encyclopedia of Tea* that "Lu Yu wrote the first book about tea—*The Classic of Tea* that benefits the scientists of agriculture in China and the related people all over the world" "No one can deny the lofty position of Lu Yu".

There are a lot of poems in *The Appreciation of Tea Poems in the Republic of Korea* by the distinguished tea professor Kim Myeonbae that panegyrizes Lu Yu and *The Classic of Tea*. For example, "Chengxintang is famous for great tea, especially for tea which is like bamboo shoots" "Who holds three bowls to Lu Tong and praises Lu Yu". Professor Jung Sanggu, chairman of Korean Association of Tea Ceremony, said with reverence, "Lu Yu is a great sage of tea. His masterpiece *The Classic of Tea* is unprecedented and can rarely be surpassed in the future. It is a great contribution to the history of tea culture in the world. Lu Yu is not only influential to the Chinese tea culture, but he also has a great influence upon the tea culture of the Republic of Korea and Japan so the tea lovers of the three countries all worship him as 'the Tea Sage'." Kurasawa Yukihiro who is a member of Japanese Tea Culture Association says in the book *The Research on Tea Ceremony*, "Lu Yu wrote *The Classic of Tea* which is the first book about tea in the world. Lu Yu's technology of making and boiling tea was 'Tea Ceremony' by people who were in the same age with him." "Japanese tea ceremony took the Chinese tea ceremony as his mother. He took a sea-voyage eastward after his birth and now he has grown up as a son of Chinese tea ceremony." When Takeuchi Minoru, a professor in Kyoto University and famous sinologist, inspected the relic of Lu Yu's tea cultures in Huzhou, he praised that the views of Huzhou were so beautiful. He said, "Lu Yu once wrote *The Classic of Tea* that made Huzhou well-known." Zhuang Wanfang, a professor and famous Chinese tea expert, pointed out, "Lu Yu stayed in Huzhou for about 30 years, the longest time of his life, and wrote the first book about tea in the world—*The Classic of Tea* that made an immortal contribution to the people and our country." And he also wrote a poem:

The Classic of Tea is adored by the world;
The culture of Huzhou makes a large contribution.
Thousands of years of history still needs to be exploited;
So let's set up ambition and research it together.

2.4 Celebrities and Anecdotes of Tea in Huzhou in the Tang Dynasty

In the Tang Dynasty, many prestigious national politicians, writers and poets left many poems and anecdotes about tea in Huzhou.

2.4.1 Yuan Gao and His World-renowned "Poem of Tea Mountain"

Yuan Gao (c. 727-c. 786), styled name Gongyi, was a minister of the Tang Dynasty and a poet. He had served successively as Shiyushi in the court, Inspector in Jingji, Inspector in Zhexi, Prefect in Shaozhou and Huzhou. He had been made an official in an important position in the ministry of supervision for many times and was awarded posthumously Minister of the Board of Rites during the period of Xianzong in the Tang Dynasty.

In the seventh year (772) of Dali Period of the Tang Dynasty, Yuan Gao patrolled Huzhou Prefecture as the inspector of Zhexi. He was hosted by Yan Zhenqing, the prefect of Huzhou at that time and was introduced to such celebrities as Lu Yu, Jiaoran, and so on. He witnessed the completion of Three Gui Pavilion, which was built in the southeast of Miaoxi Temple of Zhushan Mountain by Yan Zhenqing and named by Lu Yu. According to "Inscription of Miaoxi Temple in Zhushan Mountain" by Yan Zhenqing: "In the seventh year of Dali Period, Yan Zhenqing was the prefect of Huzhou. Yuan Gao, who was the inspector of Zhexi and Shiyushi in the court at that time, toured Huzhou Prefecture. They met each other in Zhushan Mountain and decided to build a pavilion in the southeast. Lu Yu named it 'Three Gui Pavilion' because it was set up on *guimao* day of *guihai* month of *guichou* year (Oct. 21st, 772). They put up a shed among the sweet scented osmanthus trees in the northwest. Within hundreds of steps around this shed, there are exuberance of trees with different flowers in the colors of orange, cyan and purple. Under the osmanthus trees, there was a walkway which was called the Imperial Commissioner Road, because Yuan Gao had walked on it." The respectfulness of Yan Zhenqing to Yuan Gao can be clearly read between the lines.

In the second year (781) of Jianzhong Period, Yuan Gao was appointed the prefect of Huzhou. Every year, he went to Guzhu Mountain to supervise the manufacture of the tribute tea. When he saw the scene of the toiling masses stop their normal jobs of farming and raising silkworms to produce the tribute tea day and night, he wrote "Poem of Tea Mountain" in the first year (784) of Xingyuan Period and submitted it to Dezong with Zisun Tribute Tea. According to *Anecdotes of Western Wuli*: "Yuan Gao, the prefect, submitted 3,600 *chuan* of the tribute tea with a piece of poem."

In Yuan Gao's eyes, there were no "beautiful girls' dancing" and no "pearls and songs" in the tea mountain. There were only common people suffering from the travail of making the tribute tea and sympathy in indignation. According to records of *Annotation to the Records of Stele*, there was a story about the tribute tea of Guzhu Mountain in Huzhou: "The time when Yuan Gao paid the tribute of tea with a piece of poem of counsel was the beginning point of the reduction of the tribute tea by the emperor." Yuan Gao also gained fame all over the world from "Poem of Tea Mountain". Li Jifu had written "Tablet Inscription about Yuan Gao's Poem of Tea Mountain". There are inscription relics of Yuan Gao's "Poem of Tea Mountain" carved on a sun-facing stone in clerical script at the north foot of Guzhu Mountain (also White Lamb Mountain), Changxing County, Huzhou City: "Yuan Gao, the prefect of the Tang Dynasty, acting upon the imperial edict, came to the tea mountain for the supervision of the tribute tea production and at the end of the mission stepped onto this highest hall to write a tea poem on March the 10th in the spring of the first year (784) of Xingyuan Period." The same story was also recorded in another book: *Notes of Metal and Stone of Zhedong and Zhexi Areas*.

History books claim that, "Yuan Gao is general and generous. He is admired by all because he dares to advise frankly." Shi Jie, Hu Yuan and Sun Fu, were called the "Three Men of Scholar" at the beginning of the Song Dynasty. Shi Jie composed the poem "Supervising Secretary Yuan Gao" and spoke highly of Yuan Gao for his upright and straightforward integrity in his poem.

2.4.2 Yu Di and Autographed Stone Inscriptions on Guzhu Mountain

Yu Di, style name Yunyuan, was a minister and a litterateur in the Tang

Dynasty. In the third year (808) of Yuanhe Period, he was appointed the minister of the board of public works (*Pingzhangshi*). He was given the title of Duke of Yan after his son married Emperor Xuanzong's daughter. He served as the prefect of Huzhou in the middle years (785-805) of Zhenyuan Period. According to *Annals of Huzhou Prefecture*, Yu Di repaired the banks of canals when he was the prefect of Huzhou. "It irrigated over three thousand hectares of fields and there were a good harvest of rice, fish and lotus nuts. People showed gratitude for him and changed the name of Di (荻) Canal to Di (頔) Canal."

During his magistration in Huzhou, he made friends with Jiaoran, the presider of Miaoxi Temple in Zhushan Mountain. They exchanged poems and praised each other.

Every spring, acting upon the order of the emperor, Yu Di went to the tribute tea house in Guzhu Mountain to supervise the manufacture of the tribute tea. He also autographed the stone inscription there. Volume Ⅲ of *Records of Metal and Stone of Wuxing* says:

> After the production of the tribute tea was finished, he mountaineered the top of Guzhu Mountain to fetch some spring water to try it. After he watched the inscription by Yuan Gao who was the former prefect, he carved the poem which was called "Tea Mountain" in the stone in March of the eighth year of Zhenyuan Period of the Tang Dynasty.

Among the relics of the cliffside inscriptions of the Tang Dynasty in Guzhu Mountain, there were eight scripts under Yuan Gao's inscription. The contents are the same as that were recorded in Volume Ⅲ of *Records of Metal and Stone of Wuxing*.

Yu Di was appointed the prefect of Xiangyang in the 14th year (798) of Zhenyuan Period and concurrently the provincial governor of Shannandong Dao. After he defeated Wu Shaochen's rebellion in the east, he recruited soldiers in a large scale and expanded forces. He was grossly insubordinate to the court and kept a heavy-handed control over the subordinates and consequently became the most important off-cutting power in Hannan. He showed some restraint after Emperor Xianzong's accession to the throne and put forth his strength to cut buffer regions. Later he entered the court and became the minister of public works but finally he was degraded to the prince's guest because of some guilt.

2.4.3 Cui Yuanliang and "You Are Tasting the Best Kind of Zisun Tea"

Cui Yuanliang, the prefect of Huzhou, together with Bai Juyi and Yuan Zhen, who were poets and writers, were Jinshi in the middle years (785-805) of Zhenyuan Period of the Tang Dynasty. After they stepped into official career, they often communicated by poems and forged a profound friendship.

Cui Yuanliang, styled name Huishu, was the supervisory censor in Yuanhe Period of Emperor Xianzong. After he became the prefect of Huzhou in the third year (823) of Changqing Period, he wrote a letter to Bai Juyi after he took his office. At that time, Bai Juyi was the prefect of Hangzhou and Yuan Zhen was survey envoy of Zhexi and the prefect of Yuezhou. On receiving Cui's letter, Bai Juyi congratulated him by sending him a poem. The title is "Poem of Congratulations to Prefect Cui for His New Post in Huzhou in Vicinity of Hang and Yue as a Letter to Inform Yuan".

Why we know each other?
Because we are from the same imperial examination.
Receiving the letter I am very happy;
Old days float in my eyes.
The border of Yue envelops and swallows the blue sea.
Pavilions of Hangzhou scrape the sky.
Wuxing is small and you should yield.
Because you are the last fairy of Penglai.

Cui Yuanliang, Bai Juyi, and Yuan Zhen passed the same imperial examination, but Cui's ranking is near the end of the list. Thus Cui wrote a poem of self-mockery, two lines of which are "The mortals do not know the god's affair; should they laugh at Penglai's last immortal?" Therefore, Bai Juyi teased:

"Wuxing was small and you should yield, because you are the last fairy of Penglai."

During his tenure as a prefect in Huzhou, Cui Yuanliang usually sent Zisun Tea and Ruoxiachun Wine of Huzhou to Bai Juyi as gifts. And Bai Juyi raved about Zisun Tea and Ruoxia Wine in his poems as well: "Please send the worry-forgetting liquor of Ruoxia to the man who is fond of it in Jiangcheng" "To repay Prefect Cui in tea mountain" and "You are all tasting the best kinds of Zisun Tea". There are also such poems as "Night Sailing from Yangwu Dock into the Moon Bay—to Cui of Huzhou Prefecture" "Morning Drinking of Huzhou Wine—to Prefect Cui" "Reply to Cui Huishu (Cui Yuanliang's styled name) on December 4th".

"Night" was written after Bai Juyi was transferred to be the prefect of Suzhou from Hangzhou. He had planned to take a boat tour through Taihu Lake to the tea mountain to attend an invitational appointment with Cui Yuanliang, the prefect of Huzhou. But when the boat entered the Moon Bay of West Mountain in Dongting, Taihu Lake, Bai Juyi saw the scenery of "water mixed with the sky in the color of dark green toward evening" "the moon immersed in the cold wave" (by Bai Juyi), and was attracted by the landscape of water merged with sky and the moonlight cast on the surface of the lake. When "my eyes and ears were dazed, my body was weak" "I heard the surging under my pillow and I was made gooseflesh all over". Thus he didn't dare to cross the south of Taihu Lake to take part in the tea party organized by Prefect Cui. Bai Juyi noted in "Night" by himself that "I often envies the tour to the tea mountain in Wuxing every spring. However when I pulled into Taihu Lake, my feeling of envy shrank".

Taihu Lake area was crisscrossed with post houses and old roads on land in the Tang Dynasty. In addition to the natural river ways, there were also artificial pools and canals everywhere. Transportation was convenient because carriages and boats could take you everywhere you want to go to. However, it still would take three to five days to go there and back. Bai Juyi had planned many a time a tour to the tea mountain in Huzhou, but it was never achieved. "When can we meet in Pinzhou? Another year has passed. One should miss his friends even though they are in different places. Otherwise the relationship between them will lessen as higher in position and older in age." That was what he had cited in his poem "To Weizhi and Prefect Cui of Huzhou, Written in Spare Time in My County". The allusion of Pinzhou mentioned in this poem was quoted from the poem "Song of South to the Yangtze River" written by Liu Yun, Satrap of Wuxing, in the Liang Dynasty of the Southern Dynasty. It depicts "a beautiful girl picking Baipin (a kind of white flower in ponds and lakes) under the sunset in Tingzhou in the south of the Yangtze River".

On October the fifteenth of the fourth year (839) of Kaicheng Period of the Tang Dynasty, Bai Juyi wrote "Five Pavilions in Baipinzhou" at the request of Yang Hangong, Prefect of Huzhou. It says: "Two hundred steps away to the southeast of Huzhou, there is a river called Zhaxi which is adjacent to Tingzhou (also called Baipin)." It was named Baipin because Liu Yun, Satrap of Wuxing in the Liang Dynasty once wrote: "Picking Baipin in Tingzhou." In that poem, the "mirror-like lake separating hearts" means that they faced each other across Taihu Lake. Besides, Bai shared his feelings of becoming more involuntary, experiencing less love and more sorrow as a senior official.

In the early spring of the following year, the prefects of Huzhou and Changzhou received orders to go to the tea mountain to supervise the production of the tribute tea. Cui Yuanliang, Prefect of Huzhou, and Jia, Prefect of Changzhou, had a tea party at Jinghui Pavilion, which was located in Xuanjiao Ridge of Xiyan Mountain of Changcheng (which is Changxing County today), Huzhou. Bai Juyi versified "Poem on Hearing about the Night Feast of Jia from Changzhou and Cui from Huzhou at Jinghui Pavilion":

Far away I am hearing of the banquet at night,
The beauties and songs are all around.
Though two prefectures are bordered under the disk,
A spring gathering is going in front of the same lamp.
The beautiful girls are dancing for their best,
You are drinking Zisun Tea together with different tastes.
I am mocking myself by the guest window at this beautiful night,
Sleepily taking cattail pollen liquor to cure my pain.

Below the poem, there was a note: "My waist got injured for falling down from my horse, and I was advised to drink cattail pollen liquor." In this case, it seemed that Bai Juyi would have participated in that feast if he had not got hurt in the waist. (Cattail pollen is a kind of herbs. It is natured, sweet, and has an effect of promoting blood circulation and removing blood stasis.)

Bai Juyi and Yuan Zhen were in the habit of drinking tea all their life. Yuan Zhen has written a poem "One-character to Seven-character Verse · Tea"; and there are more than ten pieces of tea poems written by Bai Juyi and the poem about Jinghui Pavilion in the tea mountain is the most popular masterpiece.

2.4.4 Zhang Wengui and His Line "To Announce the Arrival of Zisun from Wuxing"

Zhang Wengui was a poet in the Tang Dynasty, and his father was Zhang Hongjing who was the minister of the Ministry of Justice during the Yuanhe Period of Emperor Xianzong. Zhang Wengui served successively as the prefect of Huzhou, the defender and observer of Guilin. He assumed the office of the prefect of Huzhou in the first year (841) of Huichang Period, and went to the tribute tea house in Guzhu Mountain to supervise the manufacture of the imperial tea each year. At that time, the tribute tea of Guzhu Mountain of Huzhou was in its heyday. According to *Annals of Wuxing* in Jiatai Period and *Annals of Huzhou Prefecture*: "Huzhou paid tribute of Zisun Tea of Guzhu Mountain every year, and it took 30,000 labourers a few months to produce it." "In the period of Huichang, Huzhou paid 18,400 *jin* of Zisun Tea as tribute to the court." After he finished his mission, Zhang Wengui carved an inscription on the rock of Laoyawo of Zhuoshe Ridge in Guzhu Mountain where there are a group of cliffside inscriptions of the Tang Dynasty. The writing is clear and recognizable and there still remains a sentence "Zhang Wengui, who lives in the east of the Yellow River, March the fourth of *guihai* year." *Guihai* is the third year (843) of Huichang Period. Zhang Wengui also wrote two poems and a sentence in Huzhou. One poem and the sentence are as follows:

(A) Three Wonders of Wuxing

Pingzhou is but a very very ordinary pond.
Even the lightness of silk can match the ramie fabric.
Under the Cool Breeze Tower, the lotus begins growing.
In the Moon Canyon, Zisun Tea germinates.
The three excellences in Wuxing are all treasures.
So I persuade you to have a visit to Wuxing.

(B) Sentence

Who can imagine that the officers and enlisted men with their flag will stay in the tea mountain of Hutouyan for so long?

2.4.5 Du Mu and "Praising Zisun Tea as King of Tea"

Du Mu, style name Muzhi, was a litterateur and poet in the late Tang Dynasty. His grandfather Du You had served successively as a prime minister for emperors of Dezong, Shunzong and Xianzong and was awarded Duke of Qiguo. However, when Du Mu was still a teenager, his father died and later his family went downhill. The sudden change of the family inspired him the ambition of devoting himself to the world. In the second year (828) of Dahe Period, Du Mu passed the imperial examination and took the post as an official collator, and served successively as a member of the think tank for Shen Chuanshi, the surveillance commissioner of Jiangxi and Xuanshe, and Niu Sengru, the envoy of Huainan. He was promoted to the investigating censor, but finally resigned because of his younger brother's illness. In the second year (837) of Kaicheng Period, he came to Xuanshe (now Xuancheng in Anhui Province) to take the post of assistant general of militia again. Du Mu had admired the landscape of Huzhou, and the first time he visited Huzhou he was entertained by Yang Hangong, the prefect of Huzhou, with water sports. While watching the performance, Du Mu met a maiden who was very beautiful, so he gave some silver as a love token to her mother and made an engagement that if he didn't come back within 10 years, the girl could marry someone else. After that, he took the post of prefect in Huangzhou, Chizhou and Muzhou successively. In the fourth year (850) of

Dazhong Period, he asked Emperor Xuanzong to appoint him the prefect of Huzhou. However, when he came to Huzhou to find the girl again, 14 years had passed and the girl had got married and had children. So Du Mu felt sad and wrote a poem called "Sighing over Flowers".

> It's my fault to find the beautiful scene so late,
> There is no need to feel forlorn while the flowers are blooming.
> The fierce wind blows the flowers fade,
> The green leaves cover the branch and the fruits are hanging the full tree.

After Du Mu became the prefect of Huzhou, he invited his good friend and fellow-townsman Li Ying, who lived in Hangzhou and often entertained himself by doing sightseeing, playing the zither, reading, reciting poems and drinking, to come to Huzhou as his staff. It could be proved by Du Mu's poem "Recruiting Li Ying as a Scholar to Huzhou":

> I want to enjoy the best time but it is too late,
> I want to sing while drinking but it is out of my league.
> Mountains under gloaming look green green far away,
> The cold water of the stream is shallow, but our friendship is deep.
> Eminent men busy themselves with drinking,
> A floating life does nothing but write poems and chase fame.
> When I find the lotus is going to sprout,
> It is time I take a boat and visit you at a snow night.

In March of the fifth year (851) of Dazhong Period, accompanied by Li Ying, Du Mu obeyed the commands of the emperor to enter Guzhu Mountain to superintend the production of the tribute tea. They entered Shuikou Town by waterway and then rode horses to Guzhu Mountain. While enjoying the natural scene they recited poems. When they reached Shuikou where they had to pass at the foot of Guzhu Mountain, Du Mu was enchanted by the wonderful natural landscape of high trees by the water, a fluttering signboard of wine which was used to attract customers in the market and the scene of mandarin ducks that instantly took off from the river. He suddenly felt a strong urge for poetic creation and composed one "Four-line Poem Entering Tea Mountain through Shuikou" on the spot.

The stream which is surrounded by thick woods and high trees attracts me,
The wine shop which has a fluttering signboard interests me.
I feel a deep regret disturbing the couple of mandarin ducks,
But they are so kind to look back when they are flying.

While Li Ying was looking at a group of girls in red snuggling up to each other and looking curiously at the new prefect who came to Guzhu Mountain to superintend the production of the tribute tea, Du Mu on a horse was also greatly pleased with himself and whipped with a smile. He worked up his intense sentiments into a poem "Entering Tea Mountain through Shuikou":

While you are looking at these beautiful girls in red skirts
Snuggling up to each other and looking at us,
I am so happy that I smile on my horse,
And wave the golden whip to recite poems.

Du Mu also wrote a five-character poem "Below the Tea Mountain" when he passed through Liucun Village before entering Guzhu Mountain:

The spring breeze is the most gentle and graceful,
The sun goes west around Liucun Village.
The sun is rounded by the clouds,
A torrent of water is ringing in the dividing stream.
You can see many flowers far away the rocks,
If you played the instruments, the birds will sing with your beats.
Then I sit on the grass with a cup of wine.
There is also a beautiful lady sitting beside.

From the fifth year (770) of Dali Period of the Tang Dynasty to the fourth year (850) of Dazhong Period when Du Mu obeyed the emperor's summons to produce Zisun Tribute Tea in Huzhou, nine dynasties had passed. Added together, it was 80 years.

It was recorded that 18,400 *jin* of Zisun Tea was paid to the court as tribute during Huichang Period (841-846) of the Tang Dynasty. There were only several years apart from Huichang to the time when Du Mu served as the prefect of Huzhou and the scene of manufacturing the tribute tea was still very spectacular in Guzhu Mountain. This could be read in Du Mu's poem "Inscription to Tea Mountain":

The gorgeous view of the mountain in Dongwu tops any scenes,
The wonderful taste of the tea transcends any floral scent.
Although I am an ordinary officer,
Producing the tribute tea for the Court also needs extraordinary abilities.
I berth my boat at the end of the stream,
And look at the waving flag and the green moss in the distance.
Beautiful girls are travelling back and forth in Liucun Village,
The turbulent water rings its way down between pine mountains.
The towering mountains with numerous steps reach the clouds,
Houses are located in the broad and bright places.
I look up and hear the talking and laughing;
I gaze down and see the houses and pavilions.

Sweet spring water gushes over like glittery gold;
The scented tea buds are carved by purple jade.
On the day of offering tribute,
The horses gallop at thundering speed.
The fluttering sleeves of the dancers become wet because of the brook;
And their singing echoes in the valley.
My singing outstrips the voice of the bird,
The snow reflects the shadow of the plum.
People in different places gather here;
All intend to receive the emperor's order.
The tree shadow covers the sunshine;
The fallen flowers form piles on the path.
The landscape informs the passing March;
I climb up to the top of the mountain to comfort sorrow with a drink.
I cannot control myself on this revisit,
And I look down to find myself a part of the dust.

When Du Mu was superintending the production of the tribute tea, he often had parties with his colleagues and guests. They sang, danced and enjoyed wines and tea. Once Du Mu was ill, so he drank tea instead of wine and called himself the fairy of tea just like Lu Yu. He wrote a poem entitled "Poem Substituting Wine to Entertain Guests for Ailment on Tea Mountain in Spring":

Ten days before the Qingming Festival,
We board the boat to play music and sing songs.
The mountains are so splendid and the clouds are all white,
Beautiful girls are playing in the sparkling of the stream.
There are half blossomed flowers under sunny cloudy sky.
But who knows that the sick prefect is me,
He instead of wine has to drink tea.

In order to response to Du Mu's "Inscription to Tea Mountain", Li Ying also wrote a poem "The Song of Baking the Tribute Tea in Tea Mountain":

Thank you for your warm reception and hospitality, my friends, let me feel at home,

March with spring wind is the best time for making the tribute tea,

People follow the red flag and go into the mountain for tea.

The vermilion gates for the baking stove open early in the morning,

There are newly picked tea buds in the tea cases gradually.

They keep plucking tea on day of fog and night of dew.

But the red stamps of the government urge order by order.

No one cares the hungry tea pickers, but they all care about the tea.

Bit by bit the tea piles up,

The smell of tea after boiling is sweeter than plum flowers.

Kneading and disturbing is the sound of thundering,

And the finished tea is tribute to the emperor with a list.

Tens of thousands of people want to have a try,

The whipping sound of the riding soldiers is like lighting and thundering.

Who knows the laborers are driven to work at midnight?

Ten days' imperial road amounts four thousand *li*,

All must be covered before the Festival of Clear and Bright.

Our Emperor is wise and likes advice,

But how can he get advice if counsellors fail to do their duty?

Although there are sumptuous food and luxurious clothing in the palace,

Far from it the officers and laborers complain.

You are concerned about the sad expression of the people,

You sit by the baking stove and taste the tea.

Many a time you taste the tea and ask yourself,
What labour do we dedicate for the tender fragrant green tea?
There are songs and wine in the mountain;
The leisure rooms and big houses are all occupied by Xianjia.
There is a lot of Xianjia and a lot of wine;
And cave swiftlets are everywhere.
There are more and more worries in your face,
And I have no word to ask the silk-clothes.
The workers give long sighs during the eagerly attentive baking of the tea,
And worry about the deadline of the government.
Civilians please do not languish,
If our prefect is promoted to the prime minister.

(Xianjia: singsong girl; a female criminal or suspect who is forced to serve as a music player and prostitute in feudal time)

Li Ying was a Jinshi of the tenth year (856) of Dazhong Period, but he was a scholar without an official position when he followed Du Mu as an assistant at that time. So he was different from Du Mu in his point of view and way of treating problems. In this poem, Li Ying had several purposes. Firstly, it reflects the worries and miseries of those who are doing their most to make the tribute tea. Secondly, it reveals the hideous faces of the officials who hold red printed announcements to urge farmers to hand tea and pick tea day and night. Thirdly, it criticizes the luminous spectacle where numerous Xianjia entertain the officials by singing and playing the instruments. Besides, it portrays the helplessness (to be reluctant to socialize guests) of Du Mu, the local chief executive, and his sympathy and the feeling of guilt to the farmers. Moreover, it reflects the hardships and exertions of the messengers to escort the tribute tea to cover the four-thousand-*li* journey in ten days. Lastly, it is the poet's reflection on the reality that the advisors ignore the costly tribute tea and his illusion that "Du Mu takes the position of the prime minister". For example, only when Du Mu is prompted to the prime minister can the Wu people get out of the heavy burden and enjoy themselves. The poems of the tea mountain of Li Ying and Du Mu are true reflections of the natural landscape of Guzhu Mountain of Huzhou and the hardship of the tea farmers as well as the grand occasion of producing the tribute tea.

There still remains the inscription of Du Mu on the cliffside inscription group of the Tang Dynasty located in Waigang in Guzhu Mountain. It says: "× × the fifth year of Dazhong Period × × Prefect Du Mu from Fanchuan in answer to the summon of the court × × spring × ×." An investigation shows that Du Mu has a villa in Fanchuan and so he was called Fanchuan. Du Mu took his office as the prefect of Huzhou in July of the fourth year (850) of Dazhong Period and left the post in August of the fifth year (851) of Dazhong Period, so he was on duty for exactly one year. He moved from the official mansion to Zhaxiguan before leaving Huzhou. He wrote a poem "Four-line Poem Inscribed to the Departure from My Office to Zhaxiguan on August the Twelfth", which describes the bustling scenes in Huzhou as "every house was glutted with singing for the celebration of an autumn harvest". After his departure, he went to Chang'an to take the position of Executive Secretariat whose duty was to draft imperial edicts. Du Mu, known as Du Fanchuan, also called Xiaodu, did not get high political status and was a bit discontent because he was honest and unwilling to ingratiate himself with senior officials. He was good at writing and chanting poems. His poems show the veiled beauty with great momentum and enjoy the same reputation with that of Li Shangying. He was good at writing political essays with generous and scathing letters. In his later years, he repaired the villa in Fanchuan and lived a quiet life.

Chapter 3

Tortuous Development:

Huzhou Tea Culture in the Song and Yuan Dynasties

In the Northern and Southern Song Dynasties, tea manufacturing gradually developed on the basis of the Tang Dynasty (618-907) and the Five Dynasties (907-960). The national tea-producing areas had expanded and tea production had also increased. There were many kinds famous refined tea in all parts of the country. People in the Song Dynasty broadened the social range and diversified the forms of tea culture. It is said in history that "Tea started in the Tang Dynasty and flourished in the Song Dynasty". Tea-drinking style of the Song people was very exquisite and delicate. They paid much attention to tea quality, the fire and its length of time for brewing tea, tea-brewing method and tea-drinking effect, etc. In the Taiping Xingguo Period (976-984) of the Northern Song Dynasty, the imperial court began to set up the imperial tea factory in Jian'an (today's Jian'ou of Fujian Province) to produce Beiyuan tribute tea, resulting in a great development of dragon-phoenix cake tea since then. Emperor Huizong, Zhao Ji (1082-1135), wrote *An Exposition on Tea* in the first year (1107) of Daguan Period. With his imperiality and his advocacy of tea science and tea culture, tea affairs and activities of that time were very prosperous. But it was also during this period that the tea production center of China was gradually shifted from Huzhou to Jian'ou area with the change of the imperial court's interest in the tribute tea.

3.1 Huzhou Tea Production after the Tea-producing Center Being Shifted Southward

In the Northern and Southern Song Dynasties, tea production of Huzhou encountered the unprecedented difficulties. The annual tea production time was significantly delayed; the amount of the tribute tea began to decline; and there were even several occurrences of tribute-paying pause.

The chief reason lies in the fact that the weather gradually transferred to the cold period, which affected the production of tea. According to historical records, since the late Tang Dynasty, China began to enter the cold period in Jiangnan area (the area in the Yangtze River Delta). Up to the Song Dynasty, the average temperature was 2°C to 3°C lower than that in the prosperous period of the Tang Dynasty. In the twelfth month of the lunar year, Taihu Lake formed thick ice, on the surface of which horse-drawn carriages could travel. Due to the severe coldness, a lot of tea leaves in Huzhou froze to death, or the tea budding delayed, thus affecting the tea-picking season. To ensure that the tribute tea could be paid to the court in time before the Qingming Festival, also known as the Tomb-Sweeping Day, officials and people of Huzhou tried many ways, including making a fire around the tea tree-forested areas to get them warm and mobilizing the local people to promote tea growth by shouting "Tea is budding" while beating gongs and drums. But the role of the people is, after all, limited. It's inevitable that China's tea production center was gradually shifted from Huzhou to Jian'ou. According to *The History of Tea Culture in the Song and Yuan Dynasties*, the tribute tea production in the Song Dynasty was transferred from Guzhu in Huzhou to Jian'an in Fujian in order to ensure that the tribute tea was transported to Bianjing (the capital of the Northern Song Dynasty, today's Kaifeng in Henan Province) on time before the Qingming Festival for the rural sacrifice ceremony—the ceremony of the ancient Chinese emperors' offering sacrifices to Heaven and Earth on the outskirts —and was bestowed on the royal family members and ministers to enjoy. However, the tea trees in Yixing in Jiangsu Province and Changxing in Huzhou delayed budding due to the low temperature, which was difficult to guarantee the timely delivery of the tribute tea to Bianjing. Ouyang Xiu (1007-1072), statesman, historian, essayist, calligrapher and poet of the Song Dynasty, once wrote "in March the capital can sample and enjoy the new tea from Jian'an which is three thousand five hundred *li* away", which illustrated that tea budding in Jian'an was earlier, and also bore reversely out the fact that time delay was an important reason for Huzhou tribute tea's gradually fading out.

In the Northern and Southern Song Dynasties, the whole court found itself in the grips of luxury style, which was also a major cause of the tribute tea being shifted southward. In the area of Fujian Province, in order to cater to the preferences of the emperor and his officials, the tribute tea cakes were embossed under the guidance of a group of senior officials and ministers like Cai Xiang (1012- 1067) and Zhao Ruli, etc. with dragon and phoenix, which were extremely elegant, beautiful and colorful with a lot of such varieties as dragon-ball cake tea (龙团胜雪), dragon-phoenix cake tea (龙凤英华), jade-leaf cake tea (玉叶长春), longevity-stalactite cake tea (延年石乳) and the like. The book, *Records of Beiyuan Tribute Tea in Xuanhe Period*, written by Xiong Fan during the Song Dynasty showed that "there are more than four thousand colors in its heyday". It was a natural thing for Huzhou tea to be despised by the imperial court, for it was incomparable to Fujian tea in shape, color and flavor which gained a high reputation for its retrofits.

On the other hand, the change of political power further prompted Jian'an tea to flourish and Zisun Tea of Guzhu to decline. It was the case that there was no official of Huzhou who could win his ear in the court of the Song Dynasty. Despite the fact that there were such people as Su Shi (1037-1101) and Wang Shipeng (1112-1171) with the status of prefects in Huzhou, Su Shi stayed in Huzhou for only eight months and then he had to leave for the capital to plead guilty for political crimes. In the case of Wang Shipeng and some others, they had weak power in the court, whose political and social influences were far from those of such eleven people of the former dynasties as Yan Zhenqing, Yuan Gao, Du Mu, etc.

Of course, throughout the Northern and Southern Song Dynasties, it was a difficult period for the development of Huzhou tea business, but it did not contribute nothing. Tea produced in Huzhou was favored by the general public, even though it was no longer valued by the court. Moreover, in the Northern and Southern Song Dynasties, with the formation of tea-drinking habits, the expansion of overseas trade of tea, and the sharp increase in the demand for tea, Huzhou's tea-planting area was becoming increasingly large and the tea output was constantly on the rise as well. According to historical records, Changxing, Anji, Deqing and Wuxing under the jurisdiction of Huzhou were all tea-producing areas designated by the court at that time. There was a saying in Song's *Records of Tianchi* that in the north of Mount Mogan, located in Deqing County of Huzhou, the natives were at tea planting and all the open or vacant areas and slots were planted with tea trees.

Zisun Tea of Guzhu in Changxing was still the tribute tea in the early Northern Song Dynasty. Hu Zai (1110-1170), litterateur of the Southern Song Dynasty, said in his work *Tiaoxi Yuyin Conghua* (*The Poetic Notes Series by the Recluse of the Tiaoxi River*) that "Zisun Tea of Guzhu was offered yearly to the court as a tribute on the Qingming Festival or the Tomb-Sweeping Day, which was first sacrificed to the ancestral temples, and then bestowed on the important courtiers", and that "in the early Song Dynasty, the Kingdom of Wuyue (907-978, one of the Five Dynasties and Ten Kingdoms, founded by Qian Liu) was absorbed into the Song Dynasty, effectively ending the kingdom; and the prefecture began to pay annual tribute with one hundred *jin* (a Chinese mass unit, 1 *jin* =500 g) of Guzhu Zisun Tea and a barrel of Jinsha (Golden Sand) Spring". Although there had been occurrences of tribute pause of Guzhu Zisun Tea once, it restored its position as a tribute in the Southern Song Dynasty. According to *Annals of Wuxing* in Jiatai Period of the Southern Song Dynasty, "Guzhu ... in the valleys, mostly grew tea trees to pay annual tribute to the court".

The ruling time of the Yuan Dynasty did not last for a very long time. Tea production and its management of Huzhou, located in the south regions to the Yangtze River, were not greatly affected but there was a certain degree of development in its area and output. According to *Laws and Regulations of the Yuan Dynasty*, Xuanhui Yuan (the name of the central ministry) of the Yuan Dynasty established the central offices in charge of tea gardens in the places like Changzhou and Huzhou. "There were more than 23,000 tea farming households picking tea buds to pay the royal court as tribute in Changzhou and Huzhou."

Later seven regional offices in charge of tea gardens were set up in Wucheng, Wukang, Deqing, Changxing, Anji, Gui'an and Yixing, all of which were in Huzhou except Yixing.

The rulers of the Yuan Dynasty did not cancel Zisun Tea's status as "the tribute tea", but set up a state bureau, Mocha Yuan (the grinding tea factory) in Guzhu Mountain. Mocha Yuan was known as Gongcha Yuan (the tribute tea factory) in the Tang Dynasty, which was located in Guzhu Mountain of Shuikou Town. History shows that "in the Yuan Dynasty 2,000 *jin* of powder tea were tributed and 90 *jin* of bud tea were added ... and that in the first year (1367) of the reign of Wu 2 *jin* of new bud tea and approximately 2,884 *jin* of powder tea from grinding were added to the tribute".

According to the book *Yinshan Zhengyao* (*Principles on Proper Diet*) by Hu Sihui (a court therapist and dietitian during the reign of the Yuan Dynasty in China), in the Yuan Dynasty, there were many kinds of tea in Huzhou with such top-grade tea as pyramid-shaped cake tea—"powder tea as tribute from Huzhou" and Zisun Queshe Tea—"the new and tender steamed tea buds known as Zisun", which were quite famous nationwide.

3.2 The Unfolding Change in the Way of Tea Production

An important feature of the tea production development in the Song and Yuan Dynasties was that the main trend of tea type production during this period was transformed from cake and ball tea to loose tea. During the Tang Dynasty priority was given to the production of pressed cake and ball tea, which basically followed the old customs of the previous six dynasties. It lasted till the earlier stage of the Song Dynasty. Moreover, in some places, the production like Beiyuan tribute tea was becoming more technically sophisticated and innovative, improving the production and technology of ancient Chinese cake and ball tea to a new peak. However, despite its delicacy, the processing technology of Song's cake and ball tea was complicated and there was much trouble with boiling and drinking them.

When tea drinking became increasingly popular among more working people, there would undoubtedly be some changes in the traditional patterns of tea production.

The innovations of tea type production in the Song Dynasty were firstly to meet the needs of the majority of tea drinkers in society. The working people who joined the tea-drinkers required the tea not only to be inexpensive, but also to be easy to brew and drink. As a result, steamed green tea and steamed green powder tea, which were not easily broken and smashed while steamed, gradually flourished on the basis of the past processing technology of cake and ball teas. During the Song Dynasty, in some tea-producing areas including Changxing area which specially picked and manufactured the tribute tea in the Tang Dynasty, it was a natural thing to produce loose tea instead of cake and ball tea in order to adapt to the needs of society ever since they no longer produced the tribute tea.

In some of the literature of the Song Dynasty, the pressed tea in the shapes of cake and ball were called "compressed tea" and the steamed green and powder tea which were steamed but not broken and smashed were known as "loose tea". According to the relevant literature, in the Song Dynasty the main compressed tea production areas included Xingguo Jun (today's Yangxin County in Hubei Province), Raozhou (today's Poyang County in Jiangxi Province), Chizhou (today's Guichi District in Anhui Province), Qianzhou (today's Ganzhou City in Jiangxi Province), Yuanzhou (today's Yichun City in Jiangxi Province), Linjiang Jun (today's Qingjiang County in Jiangxi Province), Shezhou (today's She County in Anhui Province), Tanzhou (today's Changjiang City in Hunan Province), Jiangling (today's Jiangling County in Hubei Province), Yuezhou (today's Yueyang City in Hunan Province), Chenzhou (today's Yuanling County in Hunan Province), Lizhou (today's Jinshi City in Hunan Province), Guangzhou (today's Hengchuan County in Henan Province), Dingzhou (today's Changde City in Hunan Province), Zhedong and Zhexi (today's Zhejiang Province), Jian'an (today's Jian'ou City in Fujian Province), etc. Loose tea production areas were mainly in Huainan Circuit, Jinghu (in part of today's Hunan and Hubei Provinces), Guizhou (the present Zigui County in Hubei Province) and Jiangnan Circuit.

In such places as Changxing in Huzhou, although in the early Northern Song Dynasty the production of cake and ball tea was transferred to the production of loose teas, in most of the time of the Song Dynasty, the production and the production areas of compressed tea still remained more than those of loose tea. In other words, in terms of the production pattern, cake and ball teas still had a slight advantage over loose tea. It was not until the Yuan Dynasty that loose tea had significantly surpassed cake and ball teas and became the major type of tea production. *Book of Agriculture* by Wang Zhen (1290-1333), published in the mid-Yuan Dynasty, indicated that there were three kinds of tea at that time, namely, Mingcha tea, powder tea and Lacha tea. The so-called Mingcha tea refers to the bud tea or the leaf tea recorded in some history books. Powder tea is made by baking the tea sprouts to make them dry first and then grinding them into powder. As for Lacha tea, it is short for Lamiancha tea, that is, the baked tea cakes are treated with wax-like thick and sticky porridge on the surface to preserve. Actually Lacha tea is cake and ball tea. Among these three types of tea, "the most expensive one was Lacha tea" with the most extraordinary processing, "which was only offered to the imperial court as tribute and was rarely seen by ordinary people in civil society". According to *Book of Agriculture*, the actual situation was that in the Yuan Dynasty, at least before the mid-Yuan Dynasty, in most regions and ethnic groups of China, people only picked, produced and drank leaf tea and powder tea except that tribute tea still adopted a tight-pressed method at that time. Ye Ziqi living in the late Yuan Dynasty and the early Ming Dynasty pointed out, in his book *Caomuzi* (an important piece of work written in 1378 in a comprehensive note-style in the Ming Dynasty), that despite the fact Jianning's tribute tea of the Yuan Dynasty was simpler than the dragon ball tea and the phoenix cake tea of the Song Dynasty, "the powder tea of Jiangxi Province and the leaf tea throughout the country had ceased to be used in civil society".

It can also be evidenced in some tea books and some relevant agricultural books of China that in the late Song Dynasty and in the Yuan Dynasty priority was given to the production of loose tea instead of the production of traditional ball and cake teas. In the existing tea books and tea documents of the Tang and Song Dynasties, when it came to tea picking and manufacturing, only the processing technology of cake and ball tea was mentioned. However, after the Yuan Dynasty, in the books such as *Book of Agriculture* and *Gist of Farming and Sericulture* (an agricultural book written by the famous Uigur agriculturist Lu Mingshan during the

Yuan Dynasty), when it came to tea manufacturing, steamed tea and steamed green and powder tea were mainly introduced, and the picking and manufacturing methods of cake and ball tea were seldom introduced or not mentioned at all. It is very obvious that the introduction of tea-processing technology in tea books or agricultural books, to a certain extent, is a reflection of the tea type production in society at that time. For instance, *Book of Agriculture* mainly introduced steamed green tea in terms of the methods of tea picking, manufacturing and storing. It is recorded in it that "it is appropriate to pick tea early and it is best to do it before Qingming and Grain Rain, which are two of the 24 Solar Terms ... Then to steam tea right with an ancient earthen utensil; when steaming is finished, spread tea out into a thin layer in a shallow basket and then roll it in a slight way when it is still wet; and then put tea in the tea-producing device, which is made from bamboo plaiting article wrapped and covered by fire-controlling indocalamus leaves, making sure to bake it with even fire and not to make it burnt in baking". This is the earliest complete record of the picking and processing technologies of loose tea or steamed green tea in China. But in the same book, the processing technology of cake and ball tea which was a focus in the Tang and Song Dynasties is very briefly introduced in a few simple sentences without a clear indication of its production process, which shows that cake tea production was outdated at that time and it's unnecessary to give a detailed introduction of it.

However, it must be pointed out here that the "outdated situation" of cake and ball tea production directed at the main production and consumption of tea in the Han nationality areas. In fact, as a traditional or special form of tea production and consumption, cake and ball tea was not only constrained in the northwest minority areas. Therefore, the transformation of Chinese tea type production in the Song and Yuan Dynasties was an inevitable result, which was in conformity with the laws of development. This production change from cake and ball tea to loose tea doesn't mean that the old and the new have an opposing and substitute relationship. Instead, they are two parallel types different in the growth and decline of their production volumes. For instance, the production and technology of loose tea still gained much development when cake and ball tea production became dominant or was at its highest in the Northern Song Dynasty. This can be seen from Ouyang Xiu's book *Notes of Resigning and Retiring to My Hometown* published in 1067. From it we can see that in the early Song Dynasty the tribute tea factory was established in Jian'an and the ball and cake tea got a sound and smooth development. However, Zhedong and Zhexi including Huzhou saw simultaneously the climax of change to loose tea and the famous tea Rizhu Tea thus came into being as well. In the years of Renzong Period of the Song Dynasty, Cai Junmo (Cai Xiang) created "Small Dragon Ball Tea as Tribute". Ouyang Xiu held the opinion that one *jin* of Small Dragon Ball Tea was worth two *liang* (1 *liang* = 50 g) of gold; it was easy to get gold but not Small Dragon Ball Tea. At the same time, Jian'an tribute tea reached its climax because of Small Dragon Ball Tea, the loose tea production areas also extended from Zhexi to Hongzhou (today's Nanchang in Jiangxi Province) Circuit and it did not take a long time for No. 1 herbal tea Shuangjing Bud Tea came into being, which far surpassed Rizhu Tea. In other words, the development of both loose tea and cake and ball tea shows that they are at least not technically contradictory but complementary to each other and mutually promotive. Therefore, the alteration and reform of tea production in the Song and Yuan Dynasties was a natural development in line with the demands of the majority of tea consumers to simplify the procedures of tea manufacturing and to reduce the cooking and drinking procedures as well.

Throughout the Song and Yuan Dynasties, Huzhou tea production was basically at a transitional stage from cake and ball tea to loose tea. Since then, the traditional tea-processing technology and its cooking and drinking customs of ancient Chinese having experienced the tea type production reform in the Song and Yuan Dynasties found their way into the Ming and Qing Dynasties and towards the modern development.

3.3 Tea-drinking Gradually Became a Custom and an Art

The Song Dynasty was the most active era in the Chinese history of tea drinking activities. Decorated tea and tea competition of the time were derived from the tribute tea. The scholars and men of letters at that time enjoyed themselves by *fencha* or tea division (pouring tea into each tea cup). The folk teahouses and restaurants were rich and colorful with their ways of tea drinking.

The most typical place of folk tea drinking in the Song Dynasty was in Lin'an (today's Hangzhou in Zhejiang Province) in the Southern Song Dynasty. When the Southern Song Dynasty established its capital here, Lin'an, as a center of teahouse culture, was becoming significant due to the exchanges and integration of the north and south tea-drinking cultures. The present teahouse (*chaguan*) was known as *chasi* in the Southern Song Dynasty. According to Volume 16 of *Records of Dreams of Past Glories* by Wu Zimu, the teahouses in Lin'an followed the arrangement style of those in Bianjing City, the capital of the Northern Song Dynasty, which were decorated with the extraordinary calligraphic works and paintings, displayed flower shelves and arranged seasonal fresh flowers. The exotic tea and tea infusion were offered for sale all the year round. In winter Qibao Leicha Tea (the ground or mashed tea with seven treasures), *sanzi* (a flour-based pastry) and Congcha Tea, etc. were sold. In the evening, they provided the tourists with movable shopping carts or stalls for tea-drinking service. At that time in the city of Lin'an, tea-drinking business was carried out day and night. Even though it was in the middle of winter with heavy snow, or it was after the middle night, there were still peddlers holding a teapot or a tea kettle to sell tea. Huzhou, located in the interior of the Hangjiahu Plain, is geographically and socially close to Hangzhou. Besides, the transportation of Huzhou has always been convenient, the economy has always been developed, and there are many wealthy businessmen, eminent masters and celebrities in Huzhou. So there is no difference between the tea-drinking custom of Huzhou and that of Hangzhou.

The teahouses in Huzhou City and towns fell into many classes in order to adapt to different consumers. Generally, the customers in the teahouses which served as tea-drinking places were mostly those from the wealthy or official families, who came here for gatherings or learning to play musical instruments, etc. The teahouses like those were referred to as "Guapaier" then. Besides, some teahouses were just in the name of tea but not in tea business, which were called "Renqing Chasi"; some were used as places where all walks of people engaged themselves in social activities and did a deal, which were known as "Shimai". Some teahouses were special for the scholar-officials to meet friends. There also existed teahouses known as "Hua Chafang", which were actually erotic places.

Doucha (tea competition or contest) is a comparative approach to the qualities of tea, which is strongly utilitarian. It was first used in the competition of tribute tea selection, delivery and market price grade. The word "dou" meaning fight or combat has summed up the intensity of such activities. Thus tea competition is also known as "tea combat".

Some said that tea competition started in the Song Dynasty; some said that it did in the Five Dynasties; others held the idea that it originated from the Tang Dynasty on the basis of the fact that Gu Bing, a painter of the Ming Dynasty, copied *Painting of Tea Competition* by Yan Liben (about 601-673), a painter in the Tang Dynasty. However, there is no denying that tea competition flourished in the Song Dynasty. The custom of tea competition swept the country in the Song Dynasty. People showed great enthusiasm, from high officials and noble lords to common people, especially the scholars and men of letters or intellectuals.

Tea competition has very strict requirement for the relevant materials, the tea sets and the cooking methods.

✻ Tea Sets

Tea sets mainly include *chaping* (tea pots), *chazhan* (teawares of cups or bowls) and *chaxian* (tea whisk). Teapots with thin neck, bulging belly, long spout and a single handle are used for boiling water and adding water, gold and silver textures of which are preferable. Su Yu held the idea that "just as the making of a musical instrument cannot give up the choice of paulownia and the making of ink cannot do without animal glue, the making of utensils for boiling tea water cannot give up the choice of gold and silver". In terms of tea infusion quality and color, porcelain teapots should be used to boil water but the utensils made of copper, iron, lead, tin, pottery, or stone are unfavorable. Teapots are better to be small, for small ones "won't take long to boil the water to the right temperature and can make accurate water pouring and adding". As for teawares of cups or bowls, the black-glazed ones produced in Jianzhou are most famous and the wares with the partridge feather pattern are also well-known. In addition, the shape of teawares is very important and should also be elaborate. A teaware with a big mouth, an oblique wall and a slightly infolding mouth rim are easy to hold the tea froth. The choice of teawares for tea drinking and tasting is to be affected by how much the tea soup is needed. "If the ware is high but the tea is less, the color is to be shaded; if the tea is more but the ware is small, the soup is not enough to steep the tea." *Chaxian* used to stir and whisk tea infusion is made of old bamboo, the body of which should be heavy and the head of which is made of sword-back-shaped split bamboo filaments with thick roots and thin ends so that a powerful and easy operation can be achieved.

＊ Water Selection

People in the Song Dynasty also attached importance to water selection. According to *An Exposition on Tea*, "the pure, soft, sweet and clean water harbors beauty, for being soft and sweet is the nature of water, which is rare". Namely, whether it was usable for making tea should be measured by water quality and water taste. Accordingly, it was not necessary to select the world famous spring water but it was essential to select the water from the clean mountain springs and constantly-used wells. River water could not be used due to its impurity. The role of water selection in the success and failure of tea competition could not be ignored.

＊ Tea Selection

Cake tea produced in Beiyuan, Jian'an was the best for tea competition, especially the white tea of them. Tea cakes were usually coated with the oil anointing on the surface when being made. So "they take on different colors of green, yellow, purple and black". Therefore, Cai Xiang thought that to judge the quality of tea cakes "is the same as a physiognomist observes the complexion". "As rosy complexion is to people, so is white to tea cakes, in which green-white wins victory over yellow-white." At the same time, in the process of making tea cakes, other leaves might be mixed together with them, which would affect *diancha* (pouring the boiling water on tea powders for several times). Before tea competition, the tea cakes should be wrapped in paper to be hammered into pieces, and then they should be rolled immediately. When the rolling was done, they should be put in a sieve "for innumerable sifting" "which must be light and flat". Only in this way could "the perfect color of tea cakes be reached with a shiny, fixed and pasted surface".

＊ Making Paste

Making paste was the first step of tea competition, before which teawares should be warmed, making "tea uneasy to become cold and tea flavor uneasy to change". It's of great importance to master the ratio between tea and water for making tea paste in the teaware. The amount of tea to be added depends on the size of teawares. Generally, one teacup or teabowl needs two *qian* (1 *qian* =5 g) of tea powder. Then pour a proper amount of boiling water into the teaware and stir evenly, making the mixture of tea and water into tea paste with a certain concentration and viscosity. After it *diantang* is needed without any delay.

*** *Diantang***

Diantang is to pour the boiling water into the teaware, which is a key step in the process of tea competition. Attention should be paid to the flexibility of the arm holding the teapot during the operation, which is supposed to be controllable to ensure an accurate point of water dropping and to prevent the water column spouting out of the teapot mouth from forming an intermittent line of water. The pouring should be paused at once when the cup reaches a moderate level not allowing any subsequent water droplets to destroy the tea infusion surface. If the arm is uncontrollable the cup will reach the full level destroying the proper ratio between the tea cup and water, which is known as *dazhuang tang* (big and strong tea infusion indicating a gushy and rushing flow of water).

*** Tea Whisking**

Tea whisking refers to the fact that, while the boiling water is being poured into the teaware with tea powder, *chaxian* or tea whisk is used to whisk and stir the tea infusion in a rotating way to form a layer of froth. *Chaxian* should be timely and moderately operated in a priority order and with varied ranges on the basis of need so as to achieve the best effect with a variety of different images of birds and animals, flowers and plants, insects and fishes, mountains and rivers, characters and calligraphies, etc., which is not marked with the presence of water. Pouring the boiling water is almost synchronized with whisking tea. Only the coordination of two and the appropriate operation can make certain the victory of tea competition, thus creating artistic beauty of it.

The above is the process of tea competition. Then how to judge the victory and failure of tea competition? *Records of Tea* pointed out the double standards: the tea infusion color and tea froth. As for the tea infusion color, the best is the well-made milky white with fresh and tender tea. Second to it from top to bottom are the colors of real greenish white, grey, yellowish white and crimson. Tea froth should be well-distributed and stay long, which is called *yaozhan* or "biting the cup" (uneasily dissipating froth on the brewing tea). The case that tea froth dissipates fast or dissipates upon water being poured into is referred to as Yunjiao Huansan (fast-dissipating froth on the brewing tea). The contestant whose bowl or cup has the water stain (also known as water footprint) being exposed out after tea froth dissipates will be declared to be a loser. At that time the result of tea competition was probably decided on by more than one competition, which should consist of the best of three games.

This is, of course, only a visual judgment. In tea competition, tasting tea is finally involved. The quality of brewing tea in the competition will be comprehensively evaluated according to its flavor, aroma and color, which requires an appropriate operation in every step from the making, boiling, brewing to drinking of tea. Only when the above-mentioned aspects could meet the best standards were the contestants able to win tea competition. This shows that the popularization of tea competition at that time went far beyond the scope of material comforts, which had become people's cherished artistic creation.

If *doucha* (tea competition) is characterized by a strong utilitarian overtone, then *fencha* or tea division (pouring tea into each tea cup) is provided with an peaceful atmosphere of literati elegance. *Fencha* is also called *chabaixi* (tea play) and *tangxi* (tea infusion play). People who are good at *fencha* will make the best use of the pouring waterline to create many varied calligraphies and paintings, which is an aesthetic experience for both creators and viewers who enjoy the sight of the images in the cups.

3.4 Celebrities and Huzhou Tea Affairs in the Song and Yuan Dynasties

In the Northern and Southern Song Dynasties, there were not many celebrities who had association with Huzhou. But in the Yuan Dynasty, many famous calligraphers and painters emerged in Huzhou, who had created paintings on tea.

The most prominent calligrapher and painter among them was Zhao Mengfu (1254-1322), whose "Painting of Tea Competition" was best-known to the people interested in tea culture. This painting is one of the collections in Taipei, China. The painting portrays four participants in two groups standing on the right and the left respectively. Each group consists of two persons with one serving as the major and the other serving as the minor. In their hands are teapots, tea cups, or portable burners and the like. Next to each of them is a tea pail on the ground. The painting vividly depicts the competition activities about the contestants' tea quality and grade in the civil society of the time, and it mainly focuses on the postures of the four participants and the expressions on their faces.

Qian Xuan, a famous painter and one of the eight talented and productive people of Huzhou at that time, created two paintings on tea. One is "Painting of Lu Tong Brewing Tea" with native white color of paper, which is still available at present. In the painting, Lu Tong (about 795-835, a poet in the Tang Dynasty), wearing a gauze bonnet and white clothes, is sitting on the ground and showing the maid and the old bearded servant how to brew tea. On the upper part of the painting remains Emperor Qianlong's inscription of a poem created by Emperor Qianlong in the mid-autumn of the 50th year (1785) of Qianlong Period of the Qing Dynasty. The other is "Painting of Tao Gu Brewing Tea on a Snow Night", which is recorded in Volume 9 of *Connoisseurship: Collection in All Ages*, but its original painting is not available.

Hu Tinghui, a native painter of Huzhou, is a contemporary of Zhao Mengfu and Qian Xuan. He created "Painting of Brewing Tea under the Pine", which was recorded in *Records of Paintings in Haogu Hall* by Yao Jiheng (1647-about 1715) in the Qing Dynasty.

The famous painter Wang Meng (1308-1385), Zhao Mengfu's grandson, is good at painting the landscapes and figures, whose masterpieces include "Painting of Reclusion in Bianshan Mountain" and "Painting of Reclusion by the Huaxi Brook". His "Painting of Brewing Tea" is still available now with varied inscriptive writings on the upper part of it.

At that time, there appeared in Huzhou one eminent monk of tea culture field, also tea master of Buddhist circle—Chinghung (1272-1352). Chinghung, whose style name is Shiwu, was from Changshu (in Jiangsu Province). His secular surname was Wen. He was an eminent monk practicing Buddhism to cultivate morality in Xiawu Mountain (today's Xiamu Mountain in Miaoxi Town of Huzhou) during the Yuan Dynasty, who was the nineteenth generation patriarch of Rinzai/Linji school (one of the five major schools of Zen Buddhism's south sect).

According to the relevant academic research, the affinity between tea and Zen came into being in the middle and late Tang Dynasty. In particular, the chief purpose of Zen Buddhism is seen as discerning one's original Buddha nature through serene heart cultivation. This is the reason why monks relied more on tea to enter Zen state. They introduced tea as an ideal medium for "discussing *gongan* (public cases)" "debating Buddhist allegorical words" and "entering the context of Zen", resulting in an affinity between tea and Zen. In addition, "Puqing Law" (a system of the monks practicing Zen while growing their own food) is the main content of *Baizhang Buddist Rules*, making "Nung Chan" (meaning Farming Zen) the major form of Zen temple economy. Consequently, tea affairs are indispensable to practicing Buddhism and farming. As the abbot of Tianhu Temple and a master of Zen, Chinghung believed in tea to be holy likewise and became an example of following the tea-Zen affinity in his career of practicing asceticism.

In "Mountain Poems" composed by Chinghung, the verses on tea are easily availabe. Among them are verses describing brewing tea, drinking tea, presenting tea to Buddha, serving tea hospitably to the guests, artemisia-soup-substituting even without tea being left, planting and baking tea, and other respects from the perspective of tea. All of these with no exception vividly show that Chinghung could not do without tea in his career of practicing asceticism and that tea and Zen were two important elements of his life.

At the end of the Yuan Dynasty, Baiyun Zen Master and Taigu Zen Master, two of "Three Korean Eminent Monks" in the religious world of the Republic of Korea, traveled across the sea to Huzhou in order to increase Zen knowledge from Shiwu Chinghung and became the disciples of him. When they returned home after finishing the studying, they not only inherited and promoted the Chinese Buddhist thoughts but also created a precedent for Korean tea culture and became the founders of it.

3.5 Huzhou's Tea Books, Tea Poetry and *Qu* Poems on Tea

In the Song Dynasty, Huzhou's tea production developed further; tea drinking had become a custom; and tea services were thriving. However, among Huzhou's tea masters of the time, only Ye Qingchen composed the book *Tractate on Waters for Tea* with one volume and Zhu Fushi wrote the book *Comprehensive Work on the Tea Mountain* with six volumes. Ye Qingchen (1000-1049), whose style name was Daoqing, was a native of Changzhou County in Suzhou. He was made a scholar in the imperial academy and the director of three agencies (Sansishi). In his book was mentioned "the best tea in Wuxing of Huzhou is Zisun Tea". As far as Zhu Fushi was concerned, his biography was unknown and his book has been lost.

In the Yuan Dynasty, there were few books on tea all over China, but Huzhou's Shen Zhen composed the book *Manuscripts on the Tea Mountain* with twelve volumes. Shen Zhen (about 1363-?), whose style name is Yuanji and pseudonym Chashan Laoren, was a native of Changxing County of Huzhou. Shen Zhen lived a reclusive life in Hengyushan Temple during the late Yuan Dynasty. Shen Zhen was contented with poverty and devoted to his Buddhism cultivation. He wrote *Manuscripts on the Tea Mountain* with twelve volumes, the original of which has been lost. In the third year (1738) of Qianlong Period of the Qing Dynasty, Bao Zhen, Magistrate of Changxing County, collected more than one hundred poems by Shen Zhen and compiled it into a book, wrote a preface to it and engraved it, which was retitled as *Collected Remnant Works of the Old Man in the Tea Mountain* with two volumes which is still available today.

In the Song Dynasty, the tribute tea was mainly transformed to Dragon and Phoenix Cake Tea of Fujian Province, and Guzhu Zisun Tea of Changxing in Huzhou was stopped for a time. Even though it was resumed afterwards, it was less in the needed amount and its reputation gradually decreased. But people still cherished, missed and praised Guzhu tea due to its high quality. Many of the poems in the Song Dynasty were created to chant Guzhu Zisun Tea of Changxing. More well-known ones are three poems written by Lu You (1125-1209), a prominent poet of the Southern Song Dynasty. One of them is "Bailing Spring Water to Brew Tea with He Yuanli and Cai Jianwu in Dongding Yard".

In addition, among them are two poems of the poet Zeng Ji (1085-1166), a prominent poet of the Southern Song Dynasty. One poem of Wang Shipeng (1112-1171) survived, who served as the prefect of Huzhou. The great poet Su Dongpo (1037-1101) in the Song Dynasty, who did not stay in Huzhou for a long time, still wrote several popular odes to tea. Su Zhe (1039-1112) also composed two poems on Guzhu Tea. Prime Minister Wang Anshi (1021-1086), an outstanding statesman, essayist and poet, wrote one poem on Huzhou tea as well.

Besides, there are many poems chanting Huzhou famous tea and commemorating the friendships in virtue of tea in the Song poetry. For instance, some poems written by Wang Zhidao (1093-1169) and Yuan Shuoyou (1140-1204) were very popular at that time. What's more, many good poems chanting Guzhu Zisun Tea written by the Song poet Ge Shengzhong (1072-1144) are valuable.

In the poetic field of the Yuan Dynasty, the poems on Huzhou famous tea were rare. However, there were poems on Huzhou Zisun Tea in Yuanqu (Qu poems; the Qu form of poetry is a type of classical Chinese poetry form). Two of them available were written by Feng Zizhen (1253-1348), whose style name is Haisu.

Chapter 4

Further Prosperity:

Huzhou Tea Culture in the Ming and Qing Dynasties

The Ming Dynasty lasted 277 years, starting with the first year (1368) of Hongwu Period and ending with the 17th year (1644) of Chongzhen Period , while the Qing Dynasty lasted 268 years, starting with the first year (1644) of Shunzhi Period and ending with the third year (1911) of Xuantong Period. The 544 years of the two dynasties witnessed an unparalleled development of Huzhou tea culture.

The relative stability of society for most of the two dynasties brought Huzhou rapid economic development, rich commodities, prosperous markets, as well as obvious changes in countryside economic structure. Tea became an important source of income for local people, especially in the hills and mountains of Changxing, Wukang, Anji, Xiaofeng and West Wucheng. The first half of the 17th century, or the turn from the Ming Dynasty to the Qing Dynasty, further marked an important period of dissemination of Chinese tea all around the world.

With these important renovations, Huzhou tea economy during the period greatly changed and even remarkably developed, as compared with the Song and Yuan Dynasties, regarding its production and management, renovation of the varieties, processing technology, ways of drinking, administration, as well as theories of tea culture and tea literati.

4.1 Tea Production and Management

Huzhou used to be an important tribute tea-producing area. So it went among tea literati: "Yangxian ranked the first for people in the Tang Dynasty, while people in the Song Dynasty took Jianzhou as their favorite. Today, the two areas produce the most tribute tea."

By the end of the Yuan Dynasty, even before Zhu Yuanzhang ascended the throne, Changxing and the places around had already been taken by Geng Bingwen, one of Zhu's generals.

At the beginning of the Ming Dynasty, the tribute tea was still paid in very large amount, so as to satisfy the need of "bartering tea for horses" during war time. In the fourth year (1371) of Hongwu Period, Huzhou paid as tribute 10,611 *jin* of bud tea, 96,808 *jin* of leaf tea. In the seventh year (1374) of Hongwu Period, 10 *jin* of Guzhu Tea was added.

Since then, due to the quelling of wars and the change of the court's taste for tea, the amount of the tribute tea from Huzhou had noticeably decreased.

According to *Annals of Huzhou Prefecture* (1576) during Wanli Period of the Ming Dynasty and *Annals of Changxing County* (1673) during Kangxi Period of the Qing Dynasty, in the eighth year (1375) of Hongwu Period of the Ming Dynasty, Zhu Yuanzhang issued the order to abolish tribute tea mills set up during the Yuan Dynasty, and reduce annual tribute tea to 2 *jin* of bud tea only. However, "Produce" of *Collections of Wuxing Anecdotes* says: "Founding Emperor of our dynasty (the Ming Dynasty) favors Guzhu Tea and has ordered 32 *jin* of the tea as annual tribute. Two days before the Qingming Festival, the magistrate should supervise in person the harvesting and processing, present the tea to the imperial palace in Nanjing, and burn incense to complete the ceremony."

In his *Three Notes in Liuyan Study* (1634), Li Rihua of the Ming Dynasty also says: "During Hongwu Period, the quantity of Guzhu Tea as tribute was reduced to a little over 50 *jin*."

In the 24th year (1391) of Hongwu Period of the Ming Dynasty, Zhu

Yuanzhang "imposed annual quotas on the tribute tea from different places, with Jianning taking the greatest share. The tea would be harvested and presented to the imperial palace directly by tea growers. No specialized institution should be set up. The tea would be named accordingly Tanchun Tea, Xianchu Tea, Cichun Tea and Zisun Tea" (c.f. *Sequel to The Classic of Tea* by Lu Tingcan of the Qing Dynasty).

In the second year (1404) of Yongle Period of the Ming Dynasty, the tribute tea from Huzhou was 30 *jin* in total. This "For Nanjing Only" tribute tea was still being paid during the years of Wanli Period of the Ming Dynasty.

According to *Sketches in Jujube Orchard* (about 1650) by Tan Qian of the Qing Dynasty, "35 *jin* of bud tea from Changxing was presented to Nanjing as tribute. Once tea from Guzhu was presented, it meant tribute of Jiecha Tea was paid". In "Annual Tribute of Bud Tea" of *Laws and Regulations of the Qing Dynasty* (1690), it so records: "Annual tribute of bud tea was under the administration of the Ministry of Official Personnel Affairs at the beginning of Shunzhi Period, and was transferred to the Ministry of Rites from the seventh year on."

In the seventh year (1650) of Shunzhi Period, the Ministry of Rites notified Department of Civil Affairs of famous tea-producing provinces that "the tribute tea should be arranged for delivery 10 days after Grain Rain every year and arrive within a stipulated period of time ... That from Huzhou should arrive within 52 days". In the 23rd year (1684) of Kangxi Period, Zhejiang Province altogether paid annually 505 *jin* of bud tea as tribute, of which 32 *jin* was from Huzhou Prefecture. From this it can be seen that Huzhou tribute tea has a history of over 1,000 years, originating with Wenshan Imperial Tea in the Three Kingdoms Period, off and on until Kangxi Period of the Qing Dynasty.

Tribute tea system was in essence a heavy economic burden imposed by the emperors on tea growers. However, the strict requirements for the quality in turn encouraged innovation of the varieties and improvement of the processing technology.

Huzhou tea was planted over large areas and featured with richness in varieties. The above mentioned tribute tea mainly grew in Guzhu of Changxing. This is rightly reflected in *Elucidation on Tea* by Xu Cishu, which says: "Yao

Bodao so exclaims: In Mingyue Valley grows marvelous tea, the best of all."

The tea from Mingyue Valley was also called Jiecha Tea, as can be seen in the following poem of "In Reply to Song of Tasting Jiecha Tea with Mao Xiaoruo and Contracting for the Tea Barter" by Wang Daohui of the Ming Dynasty:

> Being a guest in Xiling at the end of last spring,
> A jar of Jiecha Tea my good friend Mao gave me.
> Attached was an ode to the tea he newly composed;
> Three hundred words in total the ode consisted of;
> In Mingyue Valley the tea is said to have been produced,
> Those from Dongshan and Miaohou are left to vie for the second.

The tea in the above poem is the tribute tea, but not Luojie Tea. A clearer account can be found in *A Book on Tea* written by Tu Benjun of the Ming Dynasty, which says: "A marvelous variety of tea grows in Guzhu, which is currently called Shuikou Tea, so as to differentiate it from Jiecha Tea."

However, during the Ming and Qing Dynasties, there truly was the sudden rise of Luojie Tea, which received much attention of the public. As a special section will be devoted to its details, suffice it to say so much in this part.

In addition, according to *Annals of Changxing County* during Kangxi Period of the Qing Dynasty, Changxing also produced Zhangwu Tea, a variety somehow inferior to Luojie Tea. "Zhangwu, a village in Pingding, a town 45 *li* northwest to the seat of the county government, produces tea that is next to Luojie Tea in quality."

Tea was planted over large areas in Changxing. At that time, high quality tea was also planted in places like Baiyan, Wuzhan, Qingdong, Xiaopu, etc. The total annual production of tea amounted to over 100,000 *jin*.

Tea was even a fixed income in the mountain areas of Anjie and Xiaofeng, which were unfavorable for raising silkworms or growing crops. According to *Annals of Xiaofeng County* during the Qing Dynasty, the best tea grew in Tianmu Mountain. Tea harvested a few days before Grain Rain was called pre-rain tea, or bud tea, which gave off the lasting aroma and was hence more valuable. At the turn from spring to summer, tea growers would go all out to harvest tea, resulting in the term Tea Harvest Season. Harvested late, the leaves would turn coarser and less aromatic, resulting in what was called dated tea. While baking the leaves, it was vital to keep an eye on the duration and temperature of the heating. Hence, quality tea should be admirable in both color and aroma. The soil to grow the tea had to be dug repeatedly throughout the year. Otherwise the produce would diminish gradually. The account shows that tea growers in Anji and Xiaofeng not only attached great importance to the tilling of tea fields but also had a sound command of the essentials of tea harvesting and baking.

During the Ming and Qing Dynasties, Wukang already had varieties of tea like Yecha Tea, Shancha Tea, Dicha Tea, Yuqian Tea, Meijian Tea, and so on. Citation from *Annals of Wukang County* during Qianlong Period in *Annals of Huzhou Prefecture* during Tongzhi Period so says: "Among the varieties of Wukang Tea, those grown in the northwest mountains are the best." "Volume V: Produce" in *Annals of Wukang County* during Daoguang Period says: "Tea grown in Tashan Mountain is of particularly good quality."

In his *Notes of Mogan Mountain*, Wu Kanghou of the Qing Dynasty expressed very high opinion of the tea grown by monks in the mountain, saying that "Grown in the mist, the tea is hence ten times more aromatic". The uniqueness of the quality is clearly shown.

In the entry of "Contemporary Names of Tea" in *Wonders of Objects* edited and annotated by Huang Yizheng of the Ming Dynasty, it is recorded that in Wucheng County there were varieties of "Wenshan Tea (grown in Wucheng) and Longposhan Tea (grown near Guzhu)", etc. From here it can be seen that Huzhou tea was rich in variety during the Ming and Qing Dynasties.

As for the harvesting of Huzhou tea, Xu Cishu so says in his *Elucidation on Tea*: "The turn from Pure Brightness (the 5th solar term) to Grain Rain (the 6th solar term) is the most favorable time for tea harvesting. Pure Brightness is too early, while Beginning of Summer (the 7th solar term) is too late. The best time is around Grain Rain."

Concerning the quality of tea, the following is an account in *Explorations into Tea* by Luo Lin: "The quality of tea lies in three aspects: color, smell and taste. In term of the color, white is the best; next comes green; yellow is undesirable. In terms of smell, aroma of the orchid is the best; aroma of broad bean flowers comes next. In terms of taste, sweetness is the best; bitterness or astringency is undesirable."

Tea leaves had to be roasted upon harvesting. Tea growers invested great care in tea roasting. Xu Cishu says in *Elucidation on Tea*: "Shortly after harvesting, the aroma of tea leaves is still kept within, which has to be brought out with fire. Being so tender, the leaves cannot endure too long a roasting. If too much is put into the pan, the leaves will not receive equal manual force. If the leaves are roasted too long in the pan, the aroma will disperse as a result of over-roasting ... At its worst, the leaves will get burned and become undrinkable all together."

Not only high requirements were attached to tea roasting, but also to the pan for the roasting: "Newly made iron pan is most undesirable, as the iron taste will completely spoil the aroma. Still more harmful than the iron taste is grease. Therefore, a pan should be reserved before hand for tea roasting only."

There were also detailed requirements for the firewood: "Only tree branches can be used as firewood for tea roasting. Tree trunks and tree leaves are forbidden. Tree trunks produce too strong a fire, while the fire produced by leaves is so mild as to easily go out. Smooth and clean the pan and roast the leaves upon harvesting. Put into the pan no more than 4 *liang* of the leaves each time. First bake the leaves soft with mild fire, and then roast with strong ones. While roasting, turn the leaves rapidly with the hand and a wooden spade until they are half done. It's ready shortly after the aroma is smelled. Move the leaves immediately with a fan into a cotton paper padded the baking box. Store the leaves in jars when the leaves are baked dry and cooled down." From here it can be seen that tea processing was extraordinarily meticulous and difficult.

To sum up, during the Ming and Qing Dynasties, Huzhou tea was not only rich in varieties and produced on a considerable scale, but also rather advanced in processing technology and managed successfully. At that time, tea was already a necessity for life and traded with good order on the market.

Generally speaking, two kinds of tea were recognized: fine and coarse. The former was also known as bud tea, while the latter, leave tea. More complicated were the classes of famed tea, which was usually classified in terms of time, place, shape and color.

In the towns, there were already tea shops. Even in the mountain areas, tea was usually managed and dealt with by "shops of tea, bamboo shoots and other local specialties". Farmers in the mountain areas even assessed their wealth according to the criterion of "an owner of 1,000 tea trees equals a lord of 1,000 tenant families" (c.f. *Annals of Wukang County* during Daoguang Period). This convincingly shows the great importance of tea at the time in the economic life of Huzhou people.

4.2 Sudden Rise of Luojie Tea

Following Wenshan Imperial Tea in the Three Kingdoms Period and Guzhu Zisun Tea in the Tang Dynasty, a new variety of Huzhou tea emerged between Jiajing Period and Wanli Period (1522-1620) of the Ming Dynasty. This was Luojie Tea. It enjoyed great popularity among the literati and tea lovers of the time, and became a unique variety of famous Jiangnan tea.

Luojie Tea was grown in Baixian Township of Changxing County in Huzhou. According to *Appraisal on Dongshan Jiecha Tea* by Zhou Gaoqi of the Ming Dynasty, "Luojie is situated a little more than 80-90 *li* south to Yixing, a hillock which marks the border between Zhejiang and Jiangsu. Changxing is to the south of the hillock. *Jie* is the local name for gully, plain between two peaks of a mountain." According to *Description of Luojie Tea* by Xiong Mingyu, "A *jie* is generally as large as 10 *li*. There are so many of them as to be difficult to count with fingers."

As a matter of fact, these gullies bordered Yixing in the north, Guangde of Anhui in the west, and were surrounded by mountains in the northeast, northwest and southwest, which were between 70-500 meters in altitude, with the 578-meter Hutong Mountain as the highest peak, while other peaks like Baishi Mountain, Xiangwang Mountain, Mingshan Mountain were all between 300-400 meters high. As a result, an alpine frigid region of some 50 square kilometers was formed. Luojie was situated at the foot of Mingshan Mountain, about 4 kilometers in length. Generally, the area behind Xiaoqinwang Temple was called Miaohoujie, while the east area was called Dongshanjie. The area was very favorable in its natural environment. According to *Appraisal on Dongshan Jiecha Tea*: "It is said there are 88 such gullies, across which flows the Daqian Stream. The water in the stream is clean and clear, moistening the roots of the tea trees."

According to *Guidelines on Jiecha Tea* by Feng Kebin, "The gullies in Dongshan face the sun in the south, bathed in the sunshine from morning to evening, moistened by the mist, and are hence able to produce tea that has a unique aroma." Similarly, *Description of Luojie Tea* says, "For tea fields in the mountains, sunshine at sunset is more important than that at sunrise. The patch behind the temple faces the west, making it a nice tea field. However, it is not as good as Dongshan, which faces the south and is particularly favored by the sunshine, making it an extraordinary tea field." As a result, the tea grown here was of unmatched quality.

Regarding excellent areas for growing Luojie Tea, a clearer description can be found in *Explorations of Luojie Tea Regions* by Zheng Gui: "There is a mountain in Luojie called Dongshan Mountain. The mountain has 4 to 5 peaks, the last of which turns to face the west, right behind the Temple of Local Land God, with a small stream separating the two. It is a mountain of sands and stones, yellow in color, with no earth." This is the so-called "best behind the temple" area, an excellent area for growing Luojie Tea.

The ancients had also long been exploring into the classes and quality of Luojie Tea. According to *Appraisal on Dongshan Jiecha Tea*, "The first class comes from behind the old temple, where one worships the local land god. The patch of field is only 2 to 3 *mu* in area, thickly covered with green grass, and shared by Yao Xiangxian of Tiaoxi and his son-in-law. The tea trees are all rather aged and produce no more than 20 *jin* of tea every year. Instead of being green, the leaves are light yellow in color. The leaf ribs are thick and pale white. After tea is processed, very little stem is left. After the tea leaves are put in boiled water, the water turns softly white in color, like dew drops, giving a sweet taste with delicate aroma, which becomes stronger when tea is sipped at." Here, a clear account of the first class Luojie Tea is given by the author, with regard to its location, ownership, amount of production, quality and features.

The author continues, "The second class (all grown on the top gullies of the mountain) comes from the patches behind the new temple, including Qipanding,

Shamaoding, Shoujingtiao, Yaobafang, and the patch possessed by Zhou family of Wujiang. Production of the tea here has to be limited in amount as well, which is delicate in aroma, white in color, lasting in taste. It looks of little difference from that grown behind the old temple, and also tastes so when sipped at. To sum up, Jiecha Tea from both places rank the best, insipid like a bamboo, delicate like a willow branch. The contemporaries think Jiecha Tea is deep in color and strong in aroma. This all roots from hearsay."

The author goes on to introduce the origins of the third and lower classes of Luojie Tea, "The third class comes from patches behind the temple, namely Zhangsha, Daguntou, Yaodong, Luodong, Wangdong, Fandong and Baishi. The fourth class (all grown on the lower gullies of the mountain) comes from patches like Xiazhangsha, Wutongdong, Yudong, Shichang, Yatoujie, Liuqingjie, Huanglong, Yanzhao and Longchi."

Non-class Jiecha Tea includes tea from patches like Changchao, Qingkou, Zhuzhuang, Guzhu and Maoshanjie.

As a native of Changxing, Zheng Gui proposed definite criteria to assess the quality of Luojie Tea in his *Explorations of Luojie Tea Regions*. "To sum up, Luojie Tea first has to be aromatic in smell, in terms of which aroma of orchid is the best, while delicate aroma of the broad bean flower comes next. The tea has to be sweet in taste, in terms of which smoothness and thickness is the best, while slight thinness comes next. It is not tea grown in Luojie that tastes bitter or astringent." This points to five criteria to assess Luojie Tea, namely "aroma, sweetness, insipidness, smoothness and thickness". The five criteria are also known as "five virtues" of Luojie Tea, which have been employed ever since by later generations.

Grain Rain (the 6th solar term) is usually the most favorable time for tea harvesting. However, the case is different for Luojie Tea. According to *Elucidation on Tea*, "People in the gullies never harvest the tea until after Beginning of Summer (the 7th solar term). Trial harvesting is called opening plantation. Tea harvested on Beginning of Summer is called spring tea. As the place is somehow colder than elsewhere, the harvest should be waited until summer, which should not be taken as something to blame."

Appraisal on Dongshan Jiecha Tea so says, "The time for harvesting and

baking Jiecha Tea is usually set three days after Beginning of Summer, which will be delayed a little if it rains." A more detailed account can be found in *Guidelines on Jiecha Tea* by Feng Kebin: "The leaves are not yet fully grown before Grain Rain, while after Beginning of Summer the stems and leaves will become too thick, which does not meet up to the standard of being fine and tender for high quality tea. The best time is at the turn of Beginning of Summer, on a fine and peaceful day, shortly after the dew disappears. Then it is the time you pick the leaves and put them into the basket. In the scorching sun, it might be so hot inside the basket as to steam. So, the basket should be covered with an umbrella until arriving at the house. Dump out all the leaves immediately, spread them thinly, and remove all the wilted, the diseased, the silk of worms, the bugs and etc., to ensure that the leaves are clean." What high demands for the job!

Zhou Gaoqi and Feng Kebin differed little in their account of the harvest time, both taking it as being around Beginning of Summer. However, in his *Description of Luojie Tea*, Xiong Mingyu says, "None is comparable with my tea from Dongshan Mountain, which is harvested five days after Grain Rain. After rapidly washed with boiled water and then brewed in the teapot for some time, the tea turns into the color of jade, which turns bright green in winter. It is sweet in taste and light in color, with a pure and lasting aroma similar to that of babies. Its drifting sweet aroma is what tea from Huqiu Hill lacks."

So it seems harvesting Luojie Tea "five days after Grain Rain" is just a trial act. Usually it is a time that harvesting is forbidden by tea growers, for "those who value tea so much would never bear hurting the trees by picking the leaves while they are still so tender" (c.f. *Elucidation on Tea*).

Besides, Jiecha Tea is sometimes "harvested again in July or August, which is called early autumn tea and of very good quality" (c.f. *Elucidation on Tea*). However, tea growers generally hate harvesting too much in case it should affect the yield in the following spring.

Roasting was already one of the frequently employed methods in processing tea during the Ming Dynasty. However, steaming was always the method employed to process Luojie Tea, i.e. steaming plus baking. Just as is accounted in *Elucidation on Tea*, "Jiecha Tea is never roasted, but steamed in the clay steamer and then baked. The reason for this is that the leaves are a little tougher as a result of late harvesting. Therefore, roasting would only burn them into fragments rather than soften them." *Guidelines on Jiecha Tea* so accounts: "Duration of steaming depends on the tenderness of the leaves, usually to the extent that the stems break and become slightly red in color. Over-done leaves will lose their freshness. Besides, the water in the steamer has to be renewed repeatedly, for boiled water might take away the taste of the tea."

The oven for tea baking was required to be maintained every year. According to *Guidelines on Jiecha Tea*, "The oven has to be maintained every year. Wet clay has to be used in the maintenance so as to add the aroma of earth. Smoke the oven with dry firewood over the night to dry its inside. Bake some coarse leaves first and bake leaves of fine quality only from the following day on. No fresh bamboo should be used for the grill, so as to avoid the invasion of bamboo smell. The leaves should be spread evenly, neither too thick nor too thin. If coal is used, remove immediately the pieces that give off smoke. It is recommendable that a fan be used to twirl the fire. Exchange in turn the position of upper grills with lower ones. Too strong a fire may burn the leaves while a too mild one may fail to bring out the desired color and luster. Non-exchange of the position of the grills may lead to uneven dryness. Only when the stems and ribs are thoroughly dried can the leaves be merged to one or two grills, and kept for the night in the highest position of the oven. Leave a tiny amount of coal in the oven to bake the leaves mildly. In the morning, the leaves are ready for storage." The hardness of the job is fully reflected in the great care.

Luojie Tea was usually stored in "new and clean porcelain jars", assisted with indocalamus leaves.

Changxing is very rich in indocalamus leaves. The Changxinggang River, which was called the Ruoxi River (the Indocalamus River) in the past, originates in the planting area of Jiecha Tea. For this reason, during the Ming and Qing Dynasties, local materials were used in storing Jiecha Tea.

Description of Luojie Tea so accounts: "In storing the tea, first weave fresh indocalamus leaves together with bamboo filaments. Envelope the jar with these leaves and put them into the jar when the tea is ready-baked and cooled down. Seal the mouth of the jar tightly and cover it with new bricks."

Wen Long of the Ming Dynasty was even more particular about the storage:

"Every time I store the tea, I would ask the woodcutter to pick some indocalamus leaves from the hill. Envelope the jar with these leaves after cleaning and drying. Then cut them into pieces with scissors and mix them with the tea. In this way, the tea will be kept fresh and green."

This was said to be a very effective way to store tea. According to *Guidelines on Jiecha Tea*, "Having gone through summer and autumn, the tea smells even more aromatic when taken out from the jar and brewed, and tastes as if fresh."

In his *Six Aspects of Jiecha Tea*, Luo Yuncheng of the Qing Dynasty recorded this special "Poem on Tea Storage":

> Pack the jar with Jiecha Tea full to the dark inwalls,
> Seal the tea with indocalamus to keep dampness out;
> Mildew in May never breeds, osmanthus scent in August never intrudes,
> It's a recipe every bright young woman knows by heart.

The poem shows that many bright women of the time were familiar with this storage of tea as well, by which the tea would be kept unaffected even "having gone through summer and autumn".

In making Jiecha Tea, the first point is to be good at choosing the water. *Elucidation on Tea* says, "High quality tea contains its aroma within, which is brought forth with water. No water, no tea." *Notes in Meihua Nunnery* also says, "Tea relies on the water to bring forth its characteristics. Proper water enhances its performance, while improper water hinders its performance." Here, the importance of water is clearly reflected.

Guidelines on Jiecha Tea believes that in Huzhou "All the water is proper that is from Jinsha Spring of Guzhu, Banyue Spring of Deqing and Guangzhu Lake of Changxing." When making tea, in addition to rinsing the tea sets with water from these sources, "tea-rinsing" is especially important for Jiecha Tea. According to *Elucidation on Tea*, "Jiecha Tea grows at the foot of the mountains, which are covered with floating sands. When washed down by the rain, the sands will settle on the tea leaves and spoil the tea to the worst if not rinsed away in the process of tea making." Therefore, "The tea leaves have to be rinsed with hot water. However, the water should not be too hot, for boiled water will bring away with it all the aroma of the tea. Hold the tea leaves into the rinsing set with a pair of bamboo sticks and rinse the leaves in the water repeatedly. Wring out the water with fingers when all the earth, wilted leaves and stems are removed. Put the leaves back into the rinsing set and cover the set. Remove the cover after a short while, and the leaves look green and give off the strong aroma. Immediately, pour boiled water over the leaves." (This is "tea-rinsing". c.f. *Guidelines on Jiecha Tea*)

Usually there are three ways of putting in the tea. "Pour the pot full of boiled water, and then put in the tea leaves. This is called top-putting. Pour the pot half with boiled water, then put in the leaves and pour the pot full. This is called mid-putting. Put the leaves at the bottom of the pot, and then pour boiled water down over the leaves. This is called bottom-putting." According to *Appraisal on Dongshan Jiecha Tea*, Jiecha Tea from Changxing "is softened after rinsing, and hence only bottom-putting is required".

In his "Postscript" to *Reflections on Exclusive Studies of Jiecha Tea*, Shen Zhou of the Ming Dynasty so appraised: "Ancient people praised plum blossoms as: giving off such an aroma that is tinged with a special pleasing quality; having such an elegance that knows no coldness. The same can be said of Jiecha Tea. Many varieties of tea may be more famous, such as Qingyuan and Wuyi from Min, Tiandi and Huqiu from Wu, Longjing from Wulin, Songluo from Xin'an, and Yunwu from Kuanglu. But none of them is comparable with Jiecha Tea." The high position of Jiecha Tea in the mind of the literati of the time is obvious to all.

Therefore, it is not strange that the famed genius Zhang Dai of Shanyin (today's Shaoxing in Zhejiang Province) should take the trouble to come down to Nanjing from Shanyin, just for Min Wenshui's tea. He was already overwhelmed with admiration after sipping at autumn Luojie Tea first, exclaiming: "The tea looks no different in color from the porcelain jar, but gives off an assailing aroma." He felt even more wonderful when next sipping at spring Luojie Tea, exclaiming: "So strong is the aroma, and so vigorous is the taste!" For this reason, he recorded these unforgettable memories in his *Memories of Dreams in Tao Nunnery*.

There was a vice attorney general of Zhejiang called Li Yulin, who was from the north. Once his good friend, Xu Ziyu from Changxing, sent him some very good Luojie Tea. Not knowing that Jiecha Tea was featured with thick stems and big leaves, he took it wrongly as something of low quality and gave it to an errand of his as an award. He became a laughing stock, for the anecdote was later recorded as a joke by Feng Mengzhen in his *Random Notes in Kuaixue Hall.*

A similar story was recorded in both *Folktales* complied by Fei Nanhui and *Random Notes in Yinxue Chamber* written by Yu Hongjian of the Qing Dynasty: "One day, before he became the prefect, Wu Xiaopao was touring Biyan with several of his friends. On their way, they passed a house in the hill, a hut with bamboo fence, tranquil, clean and distinctive. The host invited them in and entertained them with a pot of tea. The teapot was of fine porcelain. Cap removed, floating in the pot were green flower-shaped tea leaves, giving off an assailing aroma. What fine tea! They asked for more. Shortly, a maid-servant with long sideburns came in, carrying a Yixing purple clay teapot and placed it on the table. The guests sniggered at the host's rapid changing with something inferior. Presently, the host stood up, poured the tea into newly taken out small cups, and hospitably presented the tea to the guests. The guests accepted the tea and took a sip. Instantly, they were overwhelmed with a sweetness throughout the tongue, a taste that was superior to any tribute tea and beyond all the wonders described by either Lu Yu or Lu Tong. Surprised, the guests asked why. The host so answered: 'The first pot is fine all right. However, it was grown high up in the hills and harvested before Gain Rain. This second pot is truly Guzhu Tea, grown in the highest cliffs that were scarcely traversed. Every mid-spring, I would hire labors to harvest and store the tea. But never too much is harvested every year.' All the guests present were full of praise. At departure, Xiaopao asked for a little to take back home. The tea had thick stems and big leaves, never in the shape of 'one bud two blades', but tasted extremely marvelous." Here the so called "thick stems and big leaves" are nothing else but Luojie Tea.

Luojie Tea was so fashionable for a period of time during the Ming and Qing Dynasties that many local officials and literati of the time conducted deep and detailed research on it. Many of them even came up with remarkable books, such as *Description of Luojie Tea*, *Appraisal on Dongshan Jiecha Tea*, *Guidelines on Jiecha Tea* and *Elucidation on Tea*, which were mentioned before. All these show that Luojie Tea may be rightly taken as a suddenly emerged new force of tea during the two dynasties, topping all the other varieties, resulting in its far-reaching influence.

4.3 Changes in the Processing Technology and Ways of Drinking

During the Ming and Qing Dynasties, Huzhou witnessed great changes in its tea-processing technology and ways of drinking, marking an important period of further development and innovation of tea-processing technology.

During the Yuan Dynasty, tea was produced mostly in the form of powder tea. However, as the tribute tea system was chiefly inherited from the Song Dynasty and boiling continued to be the way to make tea, cake tea was still produced to a large proportion. By the beginning of the Ming Dynasty, steamed bud tea had already been very popular among the folks, which means directly baking the tea dry after steaming, without mashing and crushing. Compared with the processing methods in the past, this one was both labor-saving and capable of keeping the natural color and aroma of the tea, and hence was very popular among people of the time.

In September of the 24th year (1391) of Hongwu Period, Zhu Yuanzhang, the founding emperor of the Ming Dynasty, issued the order to "abolish cake tea and accept bud tea as the tribute tea only". From then on, steamed bud tea was replaced with the more labor-saving and more aroma catalyzing roasted bud tea, i.e. to replace steaming with roasting in tea fixation, and boiling with brewing. As a result, a time came when bud tea (or leaf tea) was solely prevailing.

Tea roasting has been mentioned in previous sections, which is also accounted in many tea books of the Ming and Qing Dynasties. For example, Xu Cishu so says in his *Elucidation on Tea*: "The current way of processing is baking on harvesting, which is capable of keeping the natural color and aroma."

In "Tea Processing" of his *Comprehensive Work on Tea*, Zhang Yuan says: "Pick out of new tea things such as wilted leaves, stems and crumbs. Use a pan

2.4 *chi* in diameter and bake 1.5 *jin* of the tea leaves each time. Put in the leaves, and roast them rapidly only when the pan is hot to the extreme. Always use strong fire and never turn it down until the leaves are fully done. Take the leaves out into a sift and roll the leaves several times. Then put the leaves back into the pan and turn the fire down gradually until the leaves are baked dry. The subtleness of the process is far beyond words. Proper fire will result in perfection of both the color and the aroma, while a slight deviation might spoil both."

Regarding the criteria for assessing tea quality, he remarks: "The secret of quality tea rests in the fineness of processing ... The starting fire determines the overall quality while the ending fire determines its clarity. Strong fire results in the aroma and clarity while a chilly pan spoils the aroma and taste. Blazing fire burns the leaves while flickering fire tarnishes its green color. Too long duration makes the leaves over-done; too short, under-done. Over-done leaves will turn yellowish; under-done, blackish. Processed in accordance with its nature, the tea will taste sweet; otherwise, astringent. While white spots matter very little, the best tea should be free from any burned spots." Others even required certain bands and strips in the leaves by rolling in the process.

These accounts reflect the highly developed tea-processing technology of the time.

In his *On Tea*, Tu Long required that the leaves should not be rolled or rubbed too hard during the roasting. Otherwise, they would be fragmented. After roasting, cool the leaves down first before storing. Otherwise, the color, aroma and taste will all be affected. Wen Long added in his *Notes on Tea*: "When roasting, it is better to have someone stand by to fan away the steaming haze so as to keep the original flavor of the tea." Many of these techniques have been in use up to today in Huzhou when roasting high quality tea.

Many other works summarized similar experiences, such as "Heat the pan beforehand. Roast with high temperatures. Roasting on harvesting. Do not roast too much each time. Do not roast too long. Be rapid in stirring the leaves. Be quick in cooling down". All these show that compared with the previous dynasty tea-processing technology in the Ming and Qing Dynasties had greatly advanced and developed both in theory and practice.

With the development of tea-processing technology, ways of tea drinking in the Ming and Qing Dynasties changed as well. After a brief coexistence of cake tea and bud tea at the beginning of the Ming Dynasty, "brewed bud tea" was rapidly popularized. This is reflected in many tea books of the two dynasties, which contain largely similar descriptions.

In this respect, we will mainly dwell on *Elucidation on Tea* by Xu Cishu. The book resulted from the writer's consultation with Yao Shaoxian, a senior expert on tea from Wuxing. Therefore, the book is in fact a summary of Wu's experience and hence rather Huzhou featured.

A quite detailed description can be found in "Tea Making" of *Elucidation on Tea*: "Prepare the tea sets before getting the water, which should be cleaned and dried. Open the sets and lay the covers upward. Lay out a porcelain pot, but do not lay it top-down on the table, for both lacquers and taste of food may spoil the tea. Hold the tea leaves in hand and put them into the teapot after the pot is filled with boiled water. Cover the pot for three breaths of time, and then pour all the water out into the jar. Put the leaves back into the pot, and shake the pot so that the aroma and color will not stagnate. Wait for another three breaths of time so that the floating stuff may get settled. Pour the tea out to entertain the guests, which is now tender and smooth as milk, sending the aroma to one's nose tip. Drinking the tea, the ill will feel noticeably better, the fatigued will feel refreshed, and the poets will feel inspired, the depressed will feel full of hope." What a detailed account of the procedures of brewing tea with a teapot and what a vivid description of the effects of tea drinking!

In order to continue to embody in "brewed bud tea" the spirit of "moral training in tea drinking" practiced in the Tang and Song Dynasties, tea literati of the Ming and Qing Dynasties proposed many very strict requirements for tea drinking. For example, regarding "situations for tea drinking", they proposed that tea must be enjoyed only in situations like "when one feels at leisure, or when one feels fatigued after composing poems, or when one feels confused in thoughts; or when one is enjoying songs or music, or when one has finished a song or a piece of music, or when one is shutting himself indoors and away from daily cares, or when one is playing a musical instrument or appreciating paintings, or when one is chatting with his intimate friends deep into the night, or when one is facing the bright window silently at the table, or when one is flirting with his bride in the nuptial chamber, or when one is receiving an honored guest or a beautiful young girl, or when one has just returned from a friend's, or on peaceful and clear days, or on slightly windy or rainy days, or on a bridge or a gayly decorated pleasure boat, or among tall trees or bamboos, or when one is appreciating flowers or playing with birds, or when one is hiding himself from the summer heat in a pavilion, or when one is burning incense in a small yard, or when the guests have taken leave after a banquet, or at one's children's home, or in a quiet and clean temple, or before a famous fountain or among queer stones," and so on. Meanwhile, they also proposed situations where tea drinking is improper.

Surely, some of these proposals are more than necessary in today's society. However, they were rather popular at that time and followed by not a few.

In "Guests to Tea" of *Elucidation on Tea*, Xu Cishu believed that "to entertain masses of friends, you may just bring out your best food and drink; to entertain new acquaintances, you may just bring out ordinary food and drink. It is only when you are chatting or arguing freely with someone of your own kind that you may have to call the lad in to make fire and prepare the tea".

For this reason, many literati set up tea houses at home, for the special purpose of entertaining honored guests and holding tea parties. This was especially popular at the end of the 16th century, i. e. in the late Ming Dynasty and the early and middle Qing Dynasty.

There is a special entry of "Tea House" in *On Tea* by Tu Long of the Ming Dynasty: "Set up a hut just beside the studio and appoint a lad specially for serving tea, in case the master might chat all day long with his friends or stay up late into cold nights. This is a first necessity and hence a must for a recluse."

In "Tea House" of his *Elucidation on Tea*, Xu Cishu also says: "Set up a special hut for tea outside your studio. Make sure the hut is dry, bright and draught-free. Set up two stoves beside the wall. Cover the stoves with white gauze, leaving open only one side, so that ashes should not blow about. Set up a table in front of the hut, and place on it kettles and pots in case for use. Set up another table for other utensils. Set up a shelf beside it for hanging the towels. Take the things into the house when they are needed for use. Cover the things immediately after use, in case they might be made dirty and thus affect the quality of the tea. The charcoal is better put far away from the stoves. Prepare as much charcoal as possible, for it burns more easily after long days of drying. The stoves are better placed a little away from the wall, and be cleaned of dirt frequently. After all, fire safety always takes the first consideration." As can be seen, safety problem is also mentioned in the book. Tea literati of the time took great pleasure in such an environment.

This method of "pot brewing", which is the same as the method used in current teahouses, originated in the middle of the Ming Dynasty. Compared with direct brewing, this indirect brewing was more favorable for bringing out the taste of the tea and hence enjoyed greater popularity among people. As a result, the method has been kept up and prospered up to today.

Xu Cishu believed "A pot of tea can only be brewed and served twice. It is delicious and tasty at the first brewing, sweet and mellow at the second, and almost flavorless at the third". When once discussing with Feng Kaizhi about their conceptions of tea drinking, he said, "The first brewed tea is like a young girl of 13, pretty and gentle; the second is like a young lady of 16, beautiful and mature; beginning from the third, it is like a married woman, with many children already." This comparison of tea to a girl echoes the famous saying: "Fine tea is always like beautiful girls."

With this superior method of tea drinking, they had very high requirements in all related respects. The first is "choice of water". In his *Comprehensive Work on Tea*, Zhang Yuan believed that "Tea is the soul for the water, while water is the body for the tea. The tea reveals its soul only when brewed in the right water, while the water shows its body only when brewing fine tea". Regarding the source of water, Lu Yu's account in *The Classic of Tea* was especially respected during the Ming and Qing Dynasties: "The first class water comes from the mountains, the second class from the rivers, and the third class from the wells."

Mountain springs in Huzhou have been already introduced in previous chapters. Hence, there is no need to elaborate any further. To sum up, the requirement is that the water should be good in quality, sweet, clean, fresh and clear. In other words, the water has to meet the standard of "The right source has no smell, and the right water has no aroma".

Great importance was also attached to the storage of water. For example, in his *Description of Luojie Tea*, Xiong Mingyu says: "For storing water, pebbles

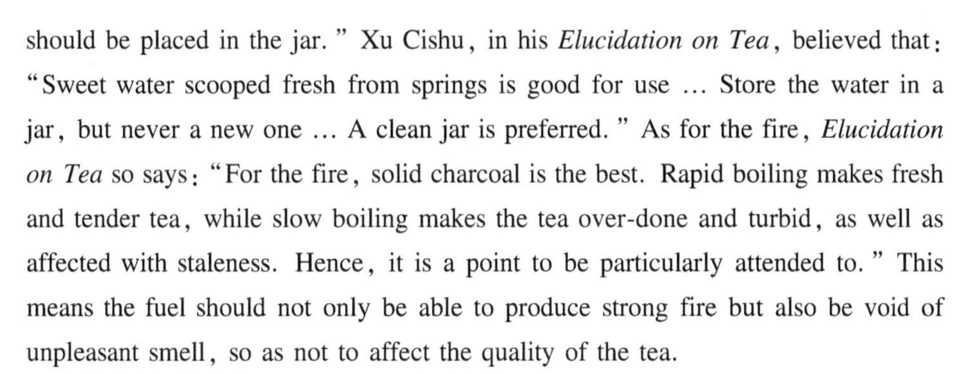

should be placed in the jar." Xu Cishu, in his *Elucidation on Tea*, believed that: "Sweet water scooped fresh from springs is good for use ... Store the water in a jar, but never a new one ... A clean jar is preferred." As for the fire, *Elucidation on Tea* so says: "For the fire, solid charcoal is the best. Rapid boiling makes fresh and tender tea, while slow boiling makes the tea over-done and turbid, as well as affected with staleness. Hence, it is a point to be particularly attended to." This means the fuel should not only be able to produce strong fire but also be void of unpleasant smell, so as not to affect the quality of the tea.

During the Ming and Qing Dynasties, great changes also occurred with regard to the utensils for making tea. First, there were a variety of stoves, such as bamboo stoves, earthen stoves, and even in-ground stoves.

Regarding the material, Zhou Gaoqi of the Ming Dynasty so says in his *Yangxian Teapots*, "Production of crumby tea and cake tea ceased in the Ming Dynasty. In this respect, the ancient people were never exceeded. However, in recent hundred years, silver, tin and porcelain from Min (largely today's Fujian) and Yu (largely today's Henan) have been replaced by pottery from Yixing in making teapots. This is the respect where today's people have exceeded the ancient." As a result, quite some famous experts on pottery and famous teapots emerged, for whom and which tea literati of the time scrambled.

As for utensils like tea kettles and teapots, Gao Lin so says in his *Eight Tips about Keeping Health*, "Porcelain and sand are the first choice for tea kettles and teapots. Copper and tin come next. Porcelain pots are the best for keeping the brewed tea, while sand kettles are the best for boiling water." However, in his *Sequel to The Classic of Tea*, Lu Tingchan of the Qing Dynasty cites this from *Daily Sketches*, "Copper tea kettles from abroad are the best for brewing tea, being made of alloy of copper and tin, and hence thin and light, fine and elegant." This shows copper tea kettles from abroad were also popular among people of the time. With regard to the color of tea bowls, as roasted bud tea was yellow-white in color after brewing, people required that "Tea bowls had better be snow-white in color. Blue-white comes next, which does no harm to the color of the tea". It can be seen that everything related to tea was highly delicate during the Ming and Qing Dynasties. Due to the changes in methods of tea processing and tea drinking, tea art was raised to a rather advanced stage.

4.4 Administration of Tea

The system of tea administration in the Ming and Qing Dynasties was largely inherited from the former dynasty, i.e. based on *Tea Code*. *Tea Code* mainly deals with three types of tea: "Official tea, which is reserved for horse trading; commercial tea, which is licensed and taxed; tribute tea, which is enjoyed by the emperor." Surveillance Commissioner appointed by the imperial court was in charge of the administration, with officials from Department of Tea Taxation and Department of Tea-horse Trade held responsible for respective aspects.

Tribute tea from Huzhou Prefecture was paid according to the quota stipulated by higher level governments. Official tea reserved for trading horses was also conducted according to the quota stipulated by governments over prefectural ones. Therefore, this part will mainly dwell on the administration of commercial tea.

According to "Tea Taxation" in *Laws and Regulations of the Ming Dynasty*, "At the beginning of the Dynasty, the quota of licensed high quality, average quality and low quality commercial tea was respectively 5,000 *jin*, 4,000 *jin*, and 3,000 *jin*. Every 7 *jin* was processed into a *bi*. The tea was freighted to the department of tea and divided equally into official tea and commercial tea. Official tea was reserved for trading horses while commercial tea was for sale. For each license of high quality, average quality and low quality tea, 100, 80 and 60 *bi* of tea was added respectively as reward. Tea smuggling was strictly banned. All commercial tea must have obtained certificates; the places that produced the tea must pay the tax; special storehouses must be built for the tea ... Special inspection offices must be established for the inspection of the tea." "As was legislated during Hongwu Period, certificates should be issued by the government and presented to prefectures and counties that produced tea. All dealers should go to the government offices to buy tea and could sell tea elsewhere only after they had paid for and obtained the certificates. Each certificate was entitled to 100 *jin* of tea. Retail certificates would be issued in cases where the amount was smaller than that for ordinary certificates. The journey would be scheduled according to the distance, with the certificate endorsed by places on the journey. Anyone could report to the government for arrest if the tea was not on the certificate or if the certificate was not with the tea. The same could be done in cases where the amount of tea was different from that on the certificate, or where someone had a surplus of tea. After sale, the original certificate should be returned to the government office in charge of tea trade. An official would be appointed for the administration by each of the related province, prefecture and county of the government."

More precise and elaborate rules are stipulated as follows: "For each certificate, 1,000 cents should be paid and 100 *jin* of tea would be entitled to. For each retail certificate, 600 cents should be paid and 60 *jin* of tea would be entitled to. Smuggling of tea would be considered as severe an offence as smuggling of salt. Buying tea from the growers in the mountains with a blanked out certificate would be considered the same as tea smuggling." "Tea growers that sell tea to dealers without the certificates would be punished for their first offence with 30 slashes, plus a fine of the original price; for their second offence, with 50 slashes; for their third offence, with 80 slashes, plus a fine of twice the original price." "After obtaining the tea, dealers must go to the inspection office to go through the inspection and have the certificates blanked out, and could leave only when clear of any smuggled goods. Offenders would be punished with 20 slashes. Falsifiers of tea certificates would be sentenced to death, plus confiscation of all the properties of the family. The tipster would be rewarded with 20 *liang* of silver."

The rigor in the administration of commercial tea and sharp distinction between reward and punishment can be easily seen. After that, the rules were amended and supplemented in the fifth year (1454) of Jingtai Period and the 31st year (1552) of Jiajing Period.

There were also legislations concerning taxation: "As was legislated during Hongwu Period, all dealers of tea should go to the taxation office to pay 1/30 tax. Both bud tea and leaf tea are taxed on the basis of approved prices, without consideration of the place of trading." Since then, amendments have been made to the taxation in different places.

Besides rigorous punishment on tea smuggling, the Ming Dynasty also made legislations concerning counterfeiting tea: "In cases of counterfeiting more than 500 *jin* of tea, both the counterfeiter and the dealer will be banished to other places. Those who belong to inland manpower sources will be banished to the frontier and forced to serve in the army. In cases of harboring more than 1,000 *jin* of tea, the shop owner would be banished as well according to the rules. Offenders of smaller amount would be convicted and punished according to the rules." It can be seen that the punishment was really rather rigorous.

As a result, in his *History of the Ming Dynasty*, Zhang Tingyu so commented on *Tea Code* of the Ming Dynasty: "The system of the Ming Dynasty was especially rigorous."

However, owing to the defects of *Tea Code* itself, together with corruption among the officials of the Ming Dynasty, especially late in the dynasty when a decline of national fortune was witnessed, malpractices occurred repeatedly in enforcing the code on official tea and commercial tea. At its worst, the maintenance of the code looked almost hopeless.

On reflection, however, compared with the Song and Yuan Dynasties, tea licensing system of the Ming Dynasty reduced the tax rate. Therefore, it was a period which witnessed a rather rapid development in tea production and tea trade.

The Qing Dynasty inherited the system of the Ming Dynasty. Huzhou Prefecture continued the practice of "licensing the trade and taxing the goods" (c.f. *Laws and Regulations of the Qing Dynasty*). The provinces differed in the annual quota of the trade. "Annual tea revenue of Zhejiang Province was 18,113.2 *liang* of silver, with a quota of 140,000 certificates, each taxed 0.1293 *liang* of silver. Taxation and licensing were practiced according to this quota." (*Laws and Regulations of the Qing Dynasty*) Similar records can be found in *Complete Laws and Regulations of the Qing Dynasty*: "Tea Taxation: All dealers should pay a certificate for every *dan* (100 *jin*) of tea from the growers, fine or coarse. Every certificate is equivalent to a taxation of 0.0033 *liang* of silver. The tea tax is supposed to be paid to the customs offices on the journey that endorse the certificate according to the rules, and is entered into the taxation part on the account book." Huzhou Prefecture must have followed this practice of Zhejiang. However, it was only the practice before the early period of the Qing Dynasty.

Since opening trade with foreign countries, especially the "opening of five trading ports in 1842", the government of the Qing Dynasty dramatically changed its tea policy. At that time, there were three major tea markets, namely Hankou,

Shanghai and Fuzhou. Tea from Zhejiang was mainly sold to Shanghai, and exported to European and American countries. In the third year (1853) of Xianfeng Period, *Lijin* System was put into practice, namely to establish customs passes on important land and water lines of communication to collect tax on passing goods, which was also known as "Li Tax" or "Lijin Tax".

Huzhou Prefecture began to enact Lijin System in the 11th year of Tongzhi Period (1872). At that time, dealers paid *lijin* tax plus local tax, the totality of which was in fact no higher than that in the Ming Dynasty. However, owing to harassment and extortion of the customs passes, dealers dumped the prices in buying tea from tea growers, so as to shuffle the tax onto tea growers. Therefore, under Lijin System, tea growers seemed increasingly exploited by commercial capital. At that time, items of tax were increased. For example, in the 20th year (1894) of Guangxu Period, 20% tax on tea was added to *lijin* tax. In 1901, after the Boxer Indemnity, a further 30% tax was added. Heavy tax plus extortion of the officials posed an extremely heavy burden to tea growers, and hence hindered the development of tea economy.

4.5 Tea Books (Articles) and Tea Literati

During the Ming and Qing Dynasties, owing to the development of tea economy, the renovation of tea varieties, as well as the changes in tea processing and drinking, tea became a focus of the national economy and the people's livelihood. Literati of the time went so far as to indulge themselves in tea drinking, and spared no efforts in writing books and articles. As a result, the prosperity of Chinese tea culture was greatly enhanced. As one of the sources of Chinese tea culture, Huzhou, for all the tests it went through throughout a history of hundreds of years, left with us quite a number of truth-provoking books on tea.

Of all these, *Elucidation on Tea* by Xu Cishu ranks among the best. Though not a native of Huzhou, he owed all his knowledge of tea to Yao Shaoxian, a native of Wuxing. Yao had a tea plantation in Guzhu, and familiarized himself with tea through years of managing the plantation. Every tea season, Xu would call at Yao's, scoop water from Jinsha Fountain or Yudou Fountain, and sip leisurely at the tea so as to assess its quality. Yao "imparted to him all the secrets about tea he learned from his own experience or elsewhere". As a result, Xu had a sound command of the mechanism of tea and was able to complete his *Elucidation on Tea*. Therefore, it should be fairly understandable to list the book as a Huzhou product.

Xu Cishu (1549-1604) was a scholar of the Ming Dynasty, native of Qiantang (currently Hangzhou), style named Ranming, pseudonymed Nanhua. He was diligent and inquisitive. Having a mania for tea, he took great pleasure in composing poems while drinking tea. With a hobby in travelling, he had a wide rang of experience and knowledge. In addition to this book, he also wrote *A Story of Merging*, etc.

Elucidation on Tea was completed in the 25th year (1597) of Wanli Period of the Ming Dynasty , consisting of one volume, about 4,700 words in total. The contents include 36 chapters, expounding such topics about tea as "tea producing" "tea processing today and in the past" "tea harvesting" "tea storing" "tea brewing", etc. In "tea harvesting" and "tea roasting", he abandons previous literature and dwells on the state of the art. In "tea processing today and in the past", he criticizes cake tea of the Song Dynasty and is against the practice of raising the price by adding spices, which spoils the original taste. He also repudiates previous statements that autumn tea was low in quality and thus unsuitable for harvesting. Besides, he gives an account of ways to fix coarse tea with steaming and fine tea with roasting, which makes it the first tea book that has ever contained an account of the roasting of green tea. The book is particularly detailed in accounting tea making and tea assessing. In addition, the account of tea ceremony on wedding and many other occasions was hitherto unknown. As a result, *Sketches in Dongcheng* so appraises the book: "It is a profound exposure of the mechanism of tea, and hence a mutual complementation to *The Classic of Tea* by Lu Yu."

The sudden rise of Luojie Tea in the Ming Dynasty raised intense interest of many literati and officials, who vied with each other to write on it. The practice lasted until the Qing Dynasty, which left us with many books on Jiecha Tea. Among these is *Description of Luojie Tea* by Xiong Mingyu, former Magistrate of

Changxing County. The book is included in *Complete Library of Chinese Ancient Tea Books*, published by Zhejiang Photographic Press in 1999, and is based on *Collection of Ancient and Modern Books*. However, *Annals of Changxing County* contains a record of *Elucidation on Luojie Tea*, which consists of 9 parts, 1,800 words in total. A comparison of the two clearly indicates that *Description of Luojie Tea* is just an excerpt of *Elucidation on Luojie Tea*.

Xiong Mingyu (about 1579-1649) was style named Liangru, pseudonymed Tanshi, native of Jinxian, Jiangxi Province. He won the Jinshi title in the 29th year of Wanli Period of the Ming Dynasty and was Magistrate of Changxing County from the 33rd year (1605) of Wanli Period for seven years. During his administration, he achieved a lot by "driving alone to investigate in the countryside and encouraging farmers to plant mulberry and tea", abandoning detrimental practices and encouraging beneficial ones. All these brought Changxing lasting peace and great prosperity. In his spare time, he completed *Elucidation on Luojie Tea*, *Elucidation on Ruoxia Wine*, and some other books on Changxing specialties. In the 43rd year (1615) of Wanli Period, he was promoted to Jishizhong to the Ministry of War Affairs. Later, he got involved in the Donglin Incident and was degraded as a result. However, he staged a comeback and retired from office as Minister of War Affairs. Changxing built the life-time temple for him when he was still alive, where inscribed on the tablet were his life events written by the famous calligrapher Dong Qichang. After death, he was listed as a distinguished official, which showed the public love and esteem for him.

Elucidation on Luojie Tea was written when Xiong Mingyu was still in office in Changxing, in about his sixth year (1611) as Magistrate of Changxing County, or, according to his own words, slightly later.

Zhou Gaoqi, the once Registrar of Changxing County, was the author of *Appraisal on Dongshan Jiecha Tea*.

Zhou Gaoqi (? -1645) was style named Bogao, native of Jiangyin, Jiangsu Province. He was very knowledgeable and majored in ancient writings. He took a hobby in tea and was good at studying teaware. He especially favored Yixing purple clay teaware, and completed in the 13th year (1640) of Chongzhen Period of the Ming Dynasty *Yangxian Teapots*, which is of great value for research on the development of Chinese ancient teawares. *Appraisal on Dongshan Jiecha Tea* was completed about the same year. The book consists of one volume, 1,500 words in total. Under sections like "Pandect" "Varieties" and "Tribute Tea", the book expounds such topics concerning Jiecha Tea as its history, origin of production, classes, processing, brewing and drinking, making it very valuable literature for successors to further research on Jiecha Tea. Currently, the book still exists in versions like *Tanji Series*, *Cuilangganguan Series*, *Lixiangbao Series*, and *Posthumous Books by Sages of Changzhou*. Besides, collected in Nanjing Library is a copy handwritten by Lu Wenchao of the Qing Dynasty, titled *Evaluation of Jiecha Tea in Dongshan Mountain*.

In addition to these, Feng Kebin, a calligrapher and painter of the Ming Dynasty, came up with his *Guidelines on Jiecha Tea* around the 15th year (1642) of Chongzhen Period.

Feng Kebin was style named Zhengqing, native of Yidu (today's Qingshan), Shandong Province. He won the title of Jinshi in the second year (1622) of Tianqi Period of the Ming Dynasty, and was once Sheriff of Huzhou. In the 4th year of Tianqi Period, he became Acting Magistrate of Changxing County, and later resigned to live in seclusion. Feng was skilled at drawing bamboo and stones, and was famed for that at the time. He named his studio "Shipu Studio" and spent his spare time drawing there. His works of bamboo and stones are inscribed and exhibited in Mumiao Pavilion in the former Wuxing Park (today's Huzhou Aishan Square). Mingxia, his concubine, was an extremely talented woman, who would serve the ink slab at his side every time he was drawing. Feng cried bitterly when Mingxia passed away. He buried her in Xianshan Mountain of Huzhou and composed this poem to her memory:

> Tears flew down every time I dreamed of this mountain,
> That ever misted my eyes when from the dream I was awoken.

His contemporaries so sighed: "Sheriff Feng is truly a spoony." His *Guidelines on Jiecha Tea* consists of about 1,000 words, including contents like "Preface" "On Tea Harvesting" "On Tea Steaming" "On Tea Baking" "On Tea Storing" "Discerning" "On Tea Making" "Assessing Spring Water" "On Teaware"

"Appropriateness of Tea Drinking" and "Inappropriateness of Tea Drinking", etc. Currently, there exists two versions of the book: *Nation-wide Collection of Books* and *Zhaodai Series*. The latter contains an appendix of five references of tea and Feng Kebin. At the end of the book is a postscript written by Yang Fuji from Zhenze in the 34th year (1695) of Kangxi Period, which so appraises the book: "Every word counts, though not so large in volume the book is." He believes that the majority of *Collected Works on Jiecha Tea* by Mao Chaomin probably found most of its source in this book.

Besides the above, another Huzhou tea book produced in the Ming Dynasty is *Tea Picking Episodes on Guzhu Mountain* by Xiao Xun.

Xiao Xun was a native of Jishui, Jiangxi Province. In the sixth year (1373) of Hongwu Period of the Ming Dynasty, he was transferred from a secretary in Ministry of Construction to a magistrate of Changxing County, and came to Guzhu Mountain to supervise the manufacturing of the tribute tea. The book was completed in the eighth year (1375) of Hongwu Period, which still exists today. However, the book contains records of tea events in the 15th year (1417) of Yongle Period and the first year (1426) of Xuande Period. Therefore, the time of book is subject to further evidence.

In addition, there is *Exclusive Studies of Jiecha Tea* by Zhou Qingshu of the Ming Dynasty, the current existence of which is, however, unknown. What exists today is just an article titled "Reflections on Exclusive Studies of Jiecha Tea" by Shen Zhou, which is cited in *Sequel to The Classic of Tea* by Lu Tingchan of the Ming Dynasty and *Authentic Manuscripts of Baishi Woodcutter* by Chen Jiru of the Ming Dynasty. According to the citations, "Qingshu lived seclusively in Changxing. He brought out teawares when I called on him and invited me to enjoy tea together. What a regret that Hongjian and Junmo never had the opportunity to meet Qingshu!" The writer's great appreciation of the book and its author should be obvious enough.

Tea was an important topic in many other books by people of the Ming Dynasty. Examples include *Anecdotes* of *Western Wuli* written by Song Gao in the 26th year (1547) of Jiaqing Period and *Anecdotes of Wuxing* written by Xu Xianzong in the 39th year (1560) of Jiaqing Period, etc. All these books give great length to Huzhou tea and Huzhou tea events.

Huzhou also produced quite a number of books on tea in the Qing Dynasty, but not so many as in the Ming Dynasty. Among them, *Collected Works on Jiecha Tea* was one of the most influential, which was complied by Mao Xiang around the 22nd year (1683) of Kangxi Period of the Qing Dynasty.

Mao Xiang (1611-1693) was a litterateur at the turn from the Ming Dynasty to the Qing Dynasty, style named Pijiang, pseudonymed Chaomin, native of Rugao, Jiangsu Province. He was already somehow talented when very young and was rather famed for that at the time. He was enrolled as Deputy Gongsheng at the end of the Ming Dynasty. Shi Kefa once recommended him to be a military inspector and later specially appointed him sheriff. He, however, never accepted the positions offered. After the Qing Dynasty began, he kept himself away from politics and took pleasure in writing. He indulged himself in hosting, feasting and touring, and was so popular at the time as to be ranked among "Four Talents in the Early Qing Dynasty". His love story with Dong Xiaowan, a well-known and unique professional female singer, was widespread and often exaggerated.

His *Collected Works on Jiecha Tea* can be found today in *Zhaodai Series*, consisting of about 1,500 words in total. The topics include origin of Jiecha Tea, its harvesting and processing, its discerning, its making and drinking, its anecdotes, etc. However, it is just a collection of excerpts from other books, 1/3 of which finds source in *Guidelines on Jiecha Tea* by Feng Kebin of the Ming Dynasty. Other sources include *Elucidation on Tea* by Xu Cishu and *Description of Luojie Tea* by Xiong Mingyu. The book contains a preface and a postscript by Zhang Chao. In "Preface", the writer illustrates the differences between tea today and tea of the past, as well as a concise and pertinent criticism to the improperness in ancient processing and drinking of tea.

There are quite some other books produced in Huzhou during the Qing Dynasty that discuss topics concerning tea. An example is *Complementary Anecdotes of Wuxing*, which was written by Zhuang Quan in the 49th year (1784) of Qianlong Period.

Among the famous mountains in Huzhou related to Huzhou tea culture, the three most representative ones are Zhushan Mountain, Guzhu Mountain and Mogan Mountain. As a result, there are more articles and poems on tea that are set in these mountains than in others. Examples include *Fu on Touring Zhushan*

Mountain by Ling Mengchu of the Ming Dynasty, *Notes on Ascending Guzhu Mountain* by You Shiren of the Ming Dynasty, and *Notes on Mogan Mountain* by Wu Kanghou of the Qing Dynasty.

A large number of tea poems were produced in Huzhou during the Ming and Qing Dynasties. According to *Huzhou Tea Poems* compiled by Zhu Nailiang in 2005, 15 tea poems were composed by Chen Ting and other 10 people of the Ming Dynasty, 26 by Jin Shengtan and other 20 people of the Qing Dynasty.

Besides tea poems, there also appeared during the Ming and Qing Dynasties a large number of antithetical couplets or other types of writings about tea. An example is a couplet in a tea pavilion in Balidian Town in the present suburbs of the city, which was written in the late Qing Dynasty and runs as follows:

> Void are all the four elements, for a moment of rest here you and me are the same;
>
> Roads run on both ends, after a cup of tea separate routes you and me may take.

Inside the century-old "Senda Teahouse" by Yifeng Bridge, there is a shop sign commonly known as "Blue Dragon Sign". The sign is a black board with words in golden lacquer on it, which are said to have been inscribed by a champion scholar native to Huzhou during Qianlong Period of the Qing Dynasty. On the board are four Chinese characters in Yan Style, vigorous and conspicuous, literally meaning "Produce Natural and National".

Chapter 5

Towards New Splendor:

Contemporary Huzhou Tea Culture

5.1 Survey of Tea Industry in Huzhou since the Republic of China

5.1.1 Tea Industry in Huzhou in the Republic of China

In the early years of the Republic of China (1912-1949), the export tea market promoted the production of tea. *New Annals of Deqing County* in 1923 recorded that "tea export increased sharply and its price was skyrocketed in recent years. The northwestern mountainous areas were almost deforested. It is better to plant tea trees than other trees". According to Zhejiang provincial tea yield released by *Zhijiang Daily* in 1915, "the yield of Xuejin Tea of Huzhou amounted to 37,423.15 *dan* (one *dan* equals to 50 kilograms) with a value of 1,217,230 silver *yuan*. The yield of Longjin Tea of Hangzhou amounted to 63,547.51 *dan* worth 3,923,660 silver *yuan*." *The Survey of Tea Industry in Hangzhou and Huzhou* in 1930 recorded that "in 1929, the tea areas in Xiaofeng and Anji Counties were 25,850 *mu*, and the annual yield of green tea was 9,900 *dan* with a value of 357,000 silver *yuan*. The tea areas in Wukang County were 5,000 *mu*. Its tea yield was added to 2,000 *dan* with a value of 60,000 silver *yuan*. The tea areas were in three villages of Nanxiang, Xixiang and Beixiang, of which Beixiang

Village was the largest tea production area and Xixiang Village was the least. The majority of teas produced in Wukang were sold in Shangbai Tea Shop or sold by peddlers in Dixi, Yuhang, Pingyao and Xianlinbu".

Tea industry experienced ups and downs in this period because of wars. Its yield rose and fell alternatively. The tea yield of Huzhou district was 12,500 *dan* in 1933, only 1/3 of that of 37,423.15 *dan* in 1915. The tea yield was 9,900 *dan* in Xiaofeng and Anji Counties in 1929, dropped to 5,040 *dan* in 1,931 and 1,900 *dan* in 1934 but rose to 9,900 *dan* in 1936.

Table 5-1 The Investigation of Tea Production in Huzhou in 1931

(Recorded by *China Industry Annals*)

Name of County	Wuxing	Changxing	Wukang	Anji	Xiaofeng	Total
Areas of Tea Plants (*mu*)	1,000	2,625	5,000	2,984	22,866	34,475
Yield (*dan*)	500	2,900	2,000	1,900	8,000	15,300
Black tea (*dan*)	—	250	100	—	—	350
Green tea (*dan*)	500	2,650	1,900	1,900	8,000	14,950

Table 5-2 Tea Yield in Huzhou in 1932 and 1933

(Recorded by *Zhejiang Economic Annals*)

(unit: *dan*)

Name of County	Deqing	Wukang	Wuxing	Changxing	Anji	Xiaofeng	Total
The year 1932	20	325	720	10,000	500	1,400	12,965
The year 1933	50	350	650	10,000	350	1,100	12,500

According to *Zhejiang Tea* in 1936, Wukang County, one of the tea areas of Hangzhou-Huzhou, produced such green tea as Qiqiang and Chaoqing (roasted fresh leaves) and black tea as well. The tea areas covered 3,484 *mu* with a yearly yield of dried tea of 4,634 *dan*, including 1,854 *dan* Qiqiang Tea. At that time Qiqiang Tea was also yielded in Dixi of Wuxing County.

During the 14 years of War of Resistance against Japanese Aggression through

1931 to 1945, tea production fell sharply with a slack tea market and desolate tea fields because most tea production areas of Huzhou were under control of Japanese. Recorded by *Zhejiang Distribution*, in 1945, tea yield in Xiaofeng was only 1,000 *dan*, and tea sales in both Anji and Xiaofeng Counties added only to 305 *dan*. An official report shows that in 1946, tea fields in Wukang County covered only 1,982 *mu*, accounting for 56.9% of that in 1936; its dried tea yield was 1,318 *dan* including 527 *dan* Qiqiang Tea, accounting for 28.4% of that in 1936. In 1949 before the founding of the People's Republic of China (PRC), tea yield in Huzhou totaled 7,170 *dan*, accounting for 46.9% of that in 1931; in the key tea areas of Anji and Xiaofeng, tea yield added to 2,570 *dan*, and tea fields were 9,168 *mu*, decreased by 80% and 41.8% respectively, compared with that in 1936. At the same time, tea yield in Changxing County was 1,752 *dan*, accounting for 17.5% of that in 1933.

5.1.2 Development of Tea Production in Huzhou after the Founding of the PRC

The tea production was soon recovered and raised after the founding of the People's Republic of China due to the recovery of the deserted tea fields, the renovation of old tea fields, the opening of new tea fields, and the popularization of technologies in the tea production under the correct leadership of the Communist Party of China and the people's government. Up to the year 1990, tea fields in Huzhou totaled 113,646 *mu* and tea yield amounted to 5,998 tons, increased by 7 times and 16.5 times respectively compared with that in 1949; and its tea value totaled over 17.7 million *yuan*, 4 times that of 1949. In the past forty years since 1949, tea production in Huzhou had made outstanding progress with accumulated 77,956 tons of tea yield, 230.546 million *yuan* of construction capital and 69.164 million *yuan* of taxes.

Tea production in Huzhou over the past fifty years since 1949 can be roughly divided into 6 periods.

In the first period (1949-1957), tea production was steadily recovered and developed. After 1949, the land ownership of the people changed production relation and liberated productive forces, motivating peasants to raise tea. Large batches of deserted tea fields were recovered and new tea fields were developed.

Such new technologies as semi-mechanical wooden tea tools were widely used, ensuring a steady growth of tea production. In 1954, tea yield reached 18,892 *dan*, more than two times that of 1949. In 1957, the tea production area amounted to 31,300 *mu*, producing tea 18,979 *dan*. Though in the late years of this period, tea yield did not remarkably increase, yet the tea field increased and the vitality of tea plants was recovered, creating conditions for future yield growth.

The second period (1958-1962) was one of stagnation and retrogression in tea production. Owing to the influence of the "Leftist" in the "Great Leap" period, some wrongdoings were taken in tea production. For example, every tea leaf should be picked and autumn and winter tea be over-picked. This was against the natural law and rule of tea production, destroying vitality of tea plants and resulting in 9.3% reduction of tea fields and 23% decrease of tea yield. As the tea tree was a perennial plant, once the tea fields were destroyed, their productivity was hard to recover. This is a hard lesson to learn. Anyway, in this period, new progress was made in tea processing, with mechanical tea factories being set up in villages, creating a new way for primary tea processing.

The third period (1963-1969) was the second recovery period of tea production. Yet as a result of the previous period, it witnessed a slow growth rate. In 1969, tea fields increased to 45,900 *mu* and tea yield reached 26,360 *dan*. Tea fields increased by 2,400 *mu* a year on average, and dried tea production increased by 1,000 *dan* a year on average. In this period, contour cultivation of tea plants, an advanced tea-planting technique, was widely used and high quality tea plants were introduced. At the same time, some professional tea villages and primary tea-processing factories were established, basically achieving primary tea processing mechanically so as to set a good beginning of a fast development of tea production.

The fourth period (1970-1982) was the period of fast growth of tea production. In the first years of this period, tea peasants were active in tea production despite the interruption from "the Cultural Revolution". They were especially motivated for tea production after the Third Plenary Session of the 11th National Congress of the Communist Party of China (CPC) in November 1978. The CPC's rural reform policies were carried out. Measures taken were such as keeping stable government tea-purchasing price, awarding tea sales to the government and reducing taxes. All these motivated the peasants so as to promote tea production. Up to the year 1982, 450 village tea farms and 300 tea factories had been set up in 500 tea production villages of 80 tea production towns. The tea production areas reached 127,735 *mu*, eight times that of the first years of the founding of the People's Republic of China. Tea fields covered an area of 100,000 *mu* with an average dried tea yield of 62.3 kilograms per *mu*. Tea yield totaled 120,783 *dan*, 17 times that of 1949. The total yield value reached 17.86 million *yuan* in Huzhou. The areas of tea fields, yield per *mu*, total yield and yield value were a high record in history. On average, more than 6,000 *mu* new tea fields added every year. Tea yield increased from 60,000 *dan* in 1978 to 120,000 *dan* in 1982. As a result, this was the fastest development period of tea production in Huzhou since 1949.

The fifth period (1983-1990) was a reform period characterized by an unsmooth road of tea production, more difficulties, hard task, sharp ups and downs but remarkable results. It could roughly be divided into three stages.

The first stage (1983-1985): Tea yield rose sharply as a result of high speed growth of tea production in the previous period. However, some kinds of tea didn't sell well because of single distribution channel, narrow market, less tea varieties, lack of market information and improperly-adjusted marketing. In 1983 and 1984, governmental tea purchasing department took such bad measures as limiting the amount of tea purchasing, stopping purchasing, lowering the price, or concealing bonus policies, forcing tea production units and peasants to limit or stop their production. As a result, tea yield dropped dramatically. Tea yield in Huzhou in 1985 was 110,980 *dan*, reduced by about 10,000 *dan* compared with that in 1982. Only 34,800 *dan* was purchased by the government, one third that of 1982. Most notably, in 1983, the sum of the value of tea purchasing, tax reduction and returns to peasants from tea factories was reduced by 5.8 million *yuan*. As a result, tea peasants' income from tea production was reduced by over 3 million *yuan*, which had greatly beaten tea peasants' motivation, resulting in reduction of tea yield in the coming three consecutive years.

The second stage (1986-1988): In this stage, policies on reforming and invigorating tea markets had been implemented. The multiple channels of tea circulation, tea export from different ports and producing a variety of tea changed tea marketing from stagnation to vitality, promoting the recovery and development of tea production in Huzhou. In 1986 and 1987, tea yield rose sharply. The year 1988 witnessed a new record in history with a total yield of 142,020 *dan*, increasing 17.6% compared with that in 1982, a previous record high in history, and with an average yield of 65.7 kg per *mu*, 3.4 kg higher than that in 1982. However, as many buyers competed in tea purchasing, there was "a tea-purchase war" in 1988. As a result, tea was purchased without strictly following the purchasing standards and tea peasants lowered the quality requirement of tea picking and tea roasting. Consequently, tea factories bought a large amount of low quality tea with high price, a potential danger for future overstock.

The third stage (1989-1990): On account of a sluggish tea market both at home and abroad, and sales stagnation of summer, autumn and low-quality tea, tea production was in a downturn with a consecutive two-year reduction of tea yield since 1989. However, because of the present multiple channels of tea circulation and production of various types of tea, tea peasants were capable of adjusting their production to the need of the market, developing brand-name and high quality tea and reducing summer and autumn tea. In so doing, market competitiveness was raised and economic benefit of tea production and tea peasants' income were increased. In these two years, despite the reduction of total yield, the increasing yield and ratio of brand-name and high quality tea brought a high economic benefit, ensuring a stable income for tea peasants. Though the total tea yield reduced by 10% in 1990 compared with that in 1989, its total value amounted to 39.18 million *yuan*, equal to that in 1989 but increased by 87.6% compared with that in 1985.

The sixth period (1991-2005) was an important period for tea production in Huzhou. It witnessed system reform of tea production and sale, the fast development of tea marketing, the adjustment of types of tea plants and varieties of tea types, the enlargement of production bases of brand-name and high quality tea, a higher degree of intensive and industrialized tea production, and the fast growth of economic benefits of the tea industry and tea peasants' income. This period can roughly be divided into two stages.

The first stage (1991-2000) was one for reform adjustment and steady development. A series of reform policies put forward by Chinese Central Government were completely implemented. The collective tea farms were run under Household Contract Responsibility System and Tea-specialized Household Contract Responsibility System, freeing tea peasants' production initiatives and liberating rural productive forces so as to motivate peasants' creativity. Generally, the areas of tea parks and tea yield were kept stable; famous and high quality tea was cultivated in a faster way and on a higher level; total yield value and economic benefits raised quickly. The areas of tea parks had been kept between 100,000 *mu* and 110,000 *mu*. Annual tea yield increased from 5,600 tons to 7,200 tons, increased by 28.6%. The total value of tea yield increased from 47 million *yuan* to 163 million *yuan*.

The second stage (2001-2005) was one for optimization and fast development. All state and collective tea farms were reformed to be run under Tea-specialized Household Contract Responsibility System. The development of famous and high quality tea speeded up, expanding their production bases, investing more money in their production, and improving their processing conditions and techniques of tea picking and frying. All these resulted in high yield and improved quality, greater market competitiveness and higher economic benefits. Especially, white tea planting areas were quickly increased and its yield and yield value were increased on a large scale. Consequently, the tea parks added from 124,755 *mu* to 168,525 *mu*, increased by 35.1%; tea yield from 7,159 tons to 8,461 tons, increased by 18.2%; its annual yield value from 164.66 million *yuan* to 220.74 million *yuan*, increased by 34.1%.

5.2 Distribution of Tea Region

Tea region in Huzhou is mainly located in northwestern part of Zhejiang Province, with the main tea farms in Anji and Changxing Counties, and in hill sections of the western part of Deqing County and Wuxing District. There were 85 tea towns including over 500 tea production villages, more than 450 village and town tea farms and 13 state tea farms. Recorded by *Tea Area Distribution in Huzhou*, there were tea farms of 128,087 *mu* producing teas of 111,159 *dan*. According to the history, present conditions, geographical conditions and tea types, tea area distribution can be divided into high quality mountain tea area and hill tea area.

5.2.1 High Quality Mountain Tea Area

Tea farms in this area are mainly distributed in the slope and valley with an altitude of 200 to 800 meters where tea plants grow in a super eco-environment and the teas are of high quality. It includes three tea areas of Hutong Mountain, Mogan Mountain and Longwang Mountain.

(1) Hutong Mountain Tea Area: It is located in Changxing County, including Baixian, Meishan, Huaikan, Shuikou, Dingjiaqiao Towns and Zhangling and Nanshan tea farms in Changchao Town. It covers a tea area of 6,413 *mu* and produced tea of 3,494 *dan*, accounting for 5% of the total tea areas and 3.1% of the total tea yield in Huzhou respectively. There are famous tea production sites such as Guzhu, Zhangling, Beichuan, Luojie, Dongshan and the adjacent mountain area for Yunwu Tea. Being near Taihu Lake, this tea area has a unique natural condition for tea growth and is rich in such famous tea in history as Zisun Tea, Luojie Tea and Dongshan Tea.

(2) Mogan Mountain Tea Area: With a suitable climate and superb ecological surroundings, Mogan Mountain has long been famous for its summer resort. This tea area includes Nanlu, Houwu, Fatou, Moganshan and Duihekou Towns in Deqing County, Qiaoxi, Meifeng Towns in Wuxing District and Kuntong Town in Anji County. Wenshan Imperial Tea, the earliest tribute tea in Zhejiang Province, was produced there and in the neighbouring Wenshan Mountain. Large sections of this area are part of Wenshan Tea Area, the most ancient one in China, dating back to the 5th century. Also produced here in history was such famous tea as Beilu Tea in Wukang and Anji Counties, Gold Lion Yunwu Tea in Qiaoxi Town, and Zifang Tea in Kuntong Town. It is also the birthplace for Mogan Huangya (yellow bud), provincial famous tea in Zhejiang. There is a tea farm, an area of 11,939 *mu* with an annual yield of 13,768 *dan*, accounting for 9.3% of the total tea areas and 12.4% of the total tea yield in Huzhou respectively. Currently, with 65% of forest coverage and filled with abundant bamboos and trees, this area has become an important production base of famous and high quality tea in Huzhou.

(3) Longwang (Dragon King) Mountain Tea Area: It is located in the

southwestern part of Anji County, including villages of Zhangcun, Baofu, Shanhe, Shanchuan, Gangkou, Hanggai, Yaocun, Panxi, Saoshe, Chiwu and Shangshu. It has a tea farm, an area of 20,510 *mu* with an annual yield of 11,830 *dan*, accounting for 16.1% of the total tea areas and 10.6% of the total tea yield in Huzhou respectively. Most part of this area has an altitude of more than 200 meters. There are a series of mountains and valleys, 78 of which have an altitude of over 1,000 meters, and Longwang Mountain has an altitude of 1,587 meters, the highest mountain in Huzhou. Its forest coverage reached 65.8%, filled with abundant bamboos and trees in the winding ranges of mountains. With fertile lands, abundant rainfall and diffuse light, high relative humidity, and covered by cloud and fog, this area is an ideal production place for white tea, a kind of nation-wide famous tea in Anji County.

5.2.2 Hill Tea Area

Under the mountain tea area and above the plains, hill tea area is located in the hill belt of Anji, Changxing and Deqing Counties and Huzhou City. It has a tea farm, an area of 89,225 *mu* with an annual yield of 82,067 *dan*, accounting for 69.7% of the total tea areas and 73.8% of the total tea yield in Huzhou respectively. So it is the major region of tea production in Huzhou. A wide hill region with an altitude of under 200 meters is covered mainly by pines, firs, bamboos and bushes. Its forest coverage is between 14% and 38%. Because of the flat slope of hills, the slope of tea farm is below 10 degrees in most cases. With a soil bed of over one meter and pH 4.5-6, its yellow mud soil is fit for tea growth. But its disadvantages are over-sticky, less fertile, only 2% of organic composition and serious lack of phosphorus and potassium. Being limestone hillock with too much calcium, some lands are unfit for growth of tea plants. Not so good as those in the mountain tea area in terms of eco-conditions, air humidity and sunlight, naturally, the tea quality in this area is lower than that of tea in the mountain area. Anyway, its advantages are as follows: convenient transportation and flat land fit for the use of advanced technology and mechanical devices and intensive planting. Additionally, there are more barren hills and land to be developed. So there is a large potential for tea production. Its structure of tea farms and production level distinguish new high productive tea area from old low productive tea area.

(1) New High Productive Tea Area: This area includes towns of Luoshe, Longshan, Wukang, Duihekou, Moganshan, Chengguan and Sanhe in Deqing; towns of Yuncao, Baique, Jingshan, Miaoxi and Nanbu, and Santianmen Tea Farm in Huzhou City; towns of Tangpu, Xiaofeng, Xiatang, Sanguan, Ancheng, Xilong, Baishuiwan, Meixi, Fengshixi, Nanbeizhuang, Lingfengsi and Nanhu Forest Farm in Anji County; towns of Baifu, Lincheng, Xiaopu in Changxing County, the County Tea Farm, Si'an and Xiaopu Forest Farm, Heping and Huicheling Tea Farm. Most of the tea farms were planted in contour and high-density cultivation pattern in the 1960s and 1970s. Only some state tea farms were planted in the way of contour cultivation in the 1950s and 1960s. Characterized by high professional planting, good infrastructure and growing management, they had a high yield with an average yield of about 100 to 150 kilograms per *mu* and a value of more than 1,000 *yuan* per *mu*. In some tea farms, the yield reached 300-350 kilograms per *mu*. This tea area mainly produces roasted green tea, with some baked green tea, flower tea, black tea and hand-made tea.

(2) Old Low Productive Tea Area: This area includes Shangbai and Sanqiao Towns in Deqing County; Dixi Town in Huzhou City; parts of the old tea areas of Dafu and Heping Towns, Niubu Village of Houyang Town in Changxing County; parts of the old tea areas of Zhangwu, Ximu and Fenghuangshan Towns, and Sanshe Village of Xiaoshi Town in Anji County. Most tea farms of this area were planted before 1949 or in the early 1950s. As a result of old tea plants, inter-cropped fields and low yield per *mu*, its economic benefits are not high. Anyway, its advantages are a long history of tea production, strong tea-making technology, especially hand-made tea and convenient transportation. Therefore, it has advantages to produce well-sold advanced hand-roasted tea and Qiqiang Tea. The early tea picking could help to win market and increase income.

5.3 Famous Huzhou Tea

As one of the tea origins in China, Huzhou is regarded as the source of famous tea by Professor Zhuang Wanfang, an authority on tea study in modern

China. There are a large number of famous tea in different dynasties. Categorized by the times, famous tea could be divided into historical traditional one and creative modern one.

5.3.1 Historical Traditional Famous Tea

5.3.1.1 Guzhu Zisun Tea

Twelve hundred years ago, this tea was in the list of the tribute tea in the Tang Dynasty. It enjoyed high popularity. Anyway, it was lost in the Qing Dynasty. However, Zisun Tea was again produced in 1978. After 1979, Zisun Tea was awarded first-class tea in four consecutive years. It was officially named Provincial Famous Tea in 1982 by Zhejiang Provincial Department of Agriculture and National Famous and High Quality Tea in Appraisal of National Famous Tea held in Nanjing in June, 1985, after being recommended as National Famous Tea in Appraisal of National Famous Tea held in Changsha City in June, 1982. It also won the Cup of High Quality Agricultural Products by the Ministry of Agriculture.

Zisun Tea has the following qualities: when it is a fresh leaf, it has bamboo-tablet shaped pointed bud and is light purple; when it is made into dried tea, its color is emerald green; when it is brewed up, its soup is grass green and crystal clear, with durable delicate flavor, mellow taste, and fat bud. Among its ingredients per 100 g tea, amino acid totals 1,897.15 mg, including 1,155 mg theanine, 149.83 mg glutamic acid, and 226 mg aspartate. Catechin reached 116.25 mg, and VC reached 223.70 mg. The content of polyphenols reached 23.49% of net weight and soluble sugar 7.88%. The content of amino acid is remarkably higher than general quality tea, VC and soluble sugar are also higher, and the other elements are equal. All these intrinsic qualities guaranteed its unique high-grade quality.

Zisun Tea is processed by the semi-roasting and drying technique, and its raw tea is high grade fresh leaves with one bud and one leaf from Guzhu, Zhangling, Zhouwujie and Nanshan in Changxing County with an area of 3,137 *mu* production base. In 1991, the yield of Zisun Tea was 8,081 kilograms.

5.3.1.2 Wenshan Imperial Tea

With a history of about 1,700 years, this tea was tribute tea in the Western and Eastern Jin and the Southern and Northern Dynasties, the earliest recorded tribute tea in Zhejiang Province and one of the earliest recorded tribute tea in China. This tea failed to be handed down a long time ago. In 1984, the agricultural department of Huzhou began to study its history and origin. Its production was recovered in 1986 and it was awarded City Famous Tea at the same year; and Provincial Excellent Famous Tea at the First and Second Famous Tea Competition held by Zhejiang Tea Association in 1987 and 1989. It is characterized by thin buds with down, grass-green, crystal clear soup, durable elegant flavor, fresh and mellow taste.

Its production places are Wenshan, Baique, Longhua, and Nanbujishan along Taihu Lake, covering an area of 1,109 *mu*. In 1991, its yield was 2,811 kilograms.

5.3.1.3 Other famous historical tea to be developed

They are Luojie Tea, Dongshan Tea in Changxing County, Qiuxianming and Bixianchun in Huzhou City, Dabeilu Tea, Xiaobeilu Tea, Dongsheng Black Tea in Deqing County, Jiumu Sweet Tea and Yongjiangxiya Tea in Anji.

5.3.2 Creative Modern Famous Tea

5.3.2.1 Anji White Tea or White Leaf Tea

This variety is a rare one with white-paper-like young bud and magnolia-like shape. The famous tea made from it has a unique style and high quality.

Recent studies show that white tea is a temperature-sensitive mutant of tea plants, a variant of genetic mutation of tea plants. Anji White Tea is a stable clonal breeding mutant. Mutation and anamorphism are important forms of evolution of organisms. Like green tea, white tea has a long history. More than 1,200 years ago, there was a record of white tea mountain in Yongjia in *The Classic of Tea* by Lu Yu. According to *An Exposition on Tea* by Zhao Ji, an emperor of the Song Dynasty, "white tea differs from ordinary tea, and is a unique one with soft and diverse branches and thin yellowish jade color. It occasionally grows in rocks where ordinary people are unable to reach. Only 4 or 5 households own it, and one of them only has one or two white tea trees, processing only 2 or 3 tea cakes ... Well processing and proper tea making procedure can ensure clear, bright yellowish tea soup, just like a white jade set in rocks, and paralleled by none." So, there was a saying that white tea is No. 1 tea. *On Dongxi Tea* by Song Zi'an in the Song

Dynasty says that there are seven kinds of tea. The first one is white tea highly valued by people. Its leaf is like a paper, regarded by people as auspicious tea. *Description of Luojie Tea* by Xiong Mingyu in the Ming Dynasty records that "I once drank certain tea, and it was white with delicate flavor and the color of this tea was precious". *Appraisal on Dongshan Jiecha Tea* by Zhou Gaoqi in the Ming Dynasty says that "white tea trees are all old plants, and the yearly yield is only about 10 kilograms. It is yellowish and its veins are whitish and thick. The tea soup is crystal-clear white and tastes sweet, with fragrance in it." These recordings show that scholars in different dynasties pay great attention to the study of white tea, regarding it as rare and the best variety of tea. Though it had long been grown since ancient times, white tea blossoms but rarely grows seeds. Its generative propagation tends to change, so the white nature of the parental tree is hard to keep stable. Consequently, being rare since ancient times, white tea has been precious and valuable.

The parental tree of white tea grew in a valley with an altitude of over 800 meters in Daxi Village, Anji County. It is a renewable tea tree with more than 100 years old, and jade-white new buds grew every spring. White tea was called by the local people Daxi White Tea. In 1958, a picture of this old tree was taken by an official of Anji Cultural Center. A paper about it was published in *People's Daily* at that time written by Zhuang Wanfang, a professor of tea study at Zhejiang University, arousing the interest of Sri Lanka breeding experts. The legendary tells that originally there were two white tea trees, one big and one small. Later, the small one died, leaving the big one living alone for more than 100 years in the valley.

In the early 1980s, a research team about selective breeding of local tea plants was organized in Huzhou, composed of 4 tea technicians from Agricultural Bureau of Huzhou City and Anji County and 2 local peasant tea technicians. The team members were Lin Shengyou, Cheng Yagu, Teng Chunying, Liu Yimin, Sheng Zhenqian and Zhang Aiping. First of all, this team investigated the resources of tea plants in Huzhou, discovering and collecting more than 90 excellent single tea plants, such as Daxi White Tea. On this basis, the team organized various research projects, beginning the selective breeding of tea plants by means of clonal breeding. After more than ten years' study and breeding, 6 clonal local tea species were successfully bred in December, 1992. Namely, they are Huzhou White Tea, Huzhou Mifeng, Mogan

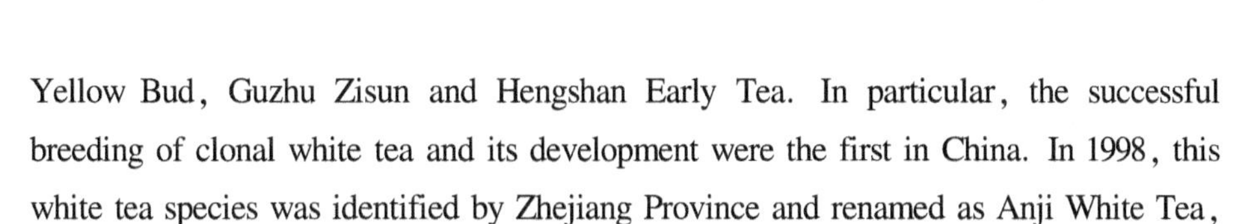

Yellow Bud, Guzhu Zisun and Hengshan Early Tea. In particular, the successful breeding of clonal white tea and its development were the first in China. In 1998, this white tea species was identified by Zhejiang Province and renamed as Anji White Tea, the only provincial endemic precious white tea in China.

The successful breeding and development of Anji White Tea plants add a rare tea species to the world. Anji White Tea processed by the young buds from the white tea trees is characterized by verdant green or golden yellow. After the tea is brewed up, it displays white leaves and green veins and is like dancing jades at the bottom of the bottle. It is fresh and has lasting flavor. The high content of various kinds of amino acid guarantees its nutritional value and health function, particularly significant for tea production. Being a rare local tea, this tea species is one for processing high grade famous tea. So, Anji White Tea has a good market and economic benefit.

Recently, with the fast development of white tea, there are tea farms with an area of over 30,000 *mu* and an annual yield of 210 tons and an annual yield value of over 200 million *yuan*. So, white tea has become a highlight of tea industry in Anji.

5.3.2.2 Mogan Yellow Bud

The first success of its planting and processing was in 1979, and was awarded by provincial and prefecture level departments in four consecutive years. And in 1982, it won the certificate of Provincial Famous Tea given by Zhejiang Provincial Agricultural Department. Its characteristics are small buds with bud down clearly seen and yellowish green. When brewed, the tea leaves dance like flying sparrows in the bottle and its soup is orange yellow and clear, and has fresh and sweet flavor. Among its ingredients, amino acid totals 4,505.22 mg, including 2,225.16 mg theanine, 565.14 mg glutamic acid, and 880.21 mg aspartate. The content of total amino acid and its sub-elements is remarkably higher than other eight kinds of famous tea tested at the same time. This tea is produced in a summer resort including Meigaohu, Hengling, Xihuding, Fushui and Bihu. It had a production area of over 300 *mu* and an yield of 225 kg in 1991.

5.3.2.3 Anji Baipian (also called Yinkeng Baipian, Yurui Tea, Daxidinggu)

The first success of its planting and processing was made in 1981 and was awarded Famous Tea at provincial and prefecture level for several times. In 1989, it was awarded National Famous Tea in the Third National Famous Tea

Assessment. It is straight but a little flat, with one leaf and one bud, and like a golden edge arrow sheath. When brewed, its soup is clear and has fresh and sweet flavor. Its leaf ends are tender, green and fat, easy to see. This tea is produced in Daxi Mountain in Anji. It had a production area of over 2,500 *mu* and an yield of over 3,000 kg in 1991.

5.3.2.4 Sangui Yuya (rain bud)

The first success of its planting and processing was in 1985 and was thus named in 1986. In 1989, it was awarded First-class Famous Tea by Zhejiang Provincial Agricultural Department. Its leaves are characterized by emerald green with down. When it is brewed up, it has chestnut aroma, thick and fresh taste; its tea soup is clear with tea leaves being yellowish green. This tea is produced in Shishan Village, Miaoxi Town, southwest of Zhushan Mountain in Huzhou City. It had a production area of 120 *mu* and an yield of over 500 kg in 1991.

5.3.2.5 Longtan Xuehao

This tea is produced by young buds of Huzhou Mifeng (Luoshe No. 1), an improved clonal tea plant in Changxing Tea Farm. It was awarded New Famous Tea and won the first prize in the Third Famous Tea Contest held by Zhejiang Tea Association and China (Hangzhou) International Tea Culture Festival in 1991. It is emerald green and has short leaves but strong buds full of down. It has a pretty shape, mellow taste and lasting aroma. The tea soup is clear with tender green leaf ends.

5.3.2.6 Zifang Queshe

This tea is produced in Zifang Village, Kuntong Town, Anji County. It was awarded Famous Tea and won the second prize in the Second Famous Tea Contest held by Zhejiang Tea Association in 1989.

5.3.2.7 Huaping Tea

This tea is produced in Huapingshan Village in Zhangcun Town, Anji County. It was awarded Famous Tea and won the second prize in the Second Famous Tea Contest held by Zhejiang Tea Association in 1989.

5.3.2.8 Yuyuan Biyu and Yuyuan Quyu

They are twin famous tea. They are produced in Nanhu Forest Farm in Anji. It was awarded Famous Tea and won the third prize in the First Famous Tea Contest held by Zhejiang Tea Association in 1987.

5.3.2.9 Hengling No. 1 Tea

This tea is made from buds and leaves of Mogan Yellow Bud (Hengling No. 1), an improved clonal local tea plant. It was awarded Famous Tea of Improved Tea Plants in 1989 in the Famous Tea of Improved Tea Plants Contest held by Zhejiang Provincial Agricultural Department. It has tender, emerald green buds with down clearly seen, and lasting aroma. The tea soup is clear with tender green leaf ends and has mellow taste.

Besides the above provincial famous tea, there is local famous tea awarded at prefecture level, namely Gucheng Qiqiang, Shanbei Lüya, Jinshi Yinhao, Jinshi Yunwu, Tianmen Biluochun, Longshan Longjin from Huzhou City; Mogan Qinglong, Mogan Yuhuachun, Mogan Jianya from Deqing County; Daxi Xiyouqing, Xuefeng Tea from Anji County.

5.4 Tea House Industry and Customs of Tea-drinking

In the Republic of China (1912-1949), tea houses in Huzhou, a wealthy city in the south of the lower reaches of the Yangtze River, were distributed in towns and countryside, along Taihu Lake, the canal, main roads, ports and remote villages. Before the 1920s, there were more than 90 tea houses in Huzhou City. *The Huzhou Monthly* recorded that there were 64 tea houses in Wuxing in 1932. Even in 1943 when Chinese people were fighting against the Japanese aggression, there were 34 tea houses. Among them, most famous tea houses were Jinguiyuan, Qiyuan and Shunyuanlou at Fumiao Temple; Zhaoyanglou, Guanfenglou, Ririsheng and Yiletian at both ends of Yifeng Bridge; Tongchunlou and Qinghelou at the South Gate of the City, Yueyanglou, Fuxinglou and Jiujianglou at the North Gate of the City; Tianyunlou, Deyilou, Zhuangyuanlou, Huifanglou and Shenggelou. There were 293 tea houses, large or small, at the early years of the People's Republic of China (PRC).

Tea houses were also flourishing in the towns where county governments were located. Before the Chinese People's War of Resistance against Japanese

Aggression, there were a population of 8,000 and 7 tea houses in Zhicheng Town of Changxing County, and the most famous one was Sanruoji. At present, the most famous one is Wanggui (forgetting to go home) Pavilion Tea House in Guzhu where Lu Yu, the Tea Sage, was lingering, and where Zisun Tribute Tea was produced, and where many celebrities at home and abroad are attracted for a tour. Guzhu Tea and Jinsha Spring are good tea and water known to people, young and old, male and female, in Changxing. Poets and literati in various dynasties came here to enjoy tea and write tea poems.

The tea houses in Old City Park in Changxing County are elegant. Sanzi Tea House in the small park at the eastern side of Renmin Square is a recreational center for the retired who bring boiled water, tea and tea sets by themselves every morning to enjoy tea and chatting.

Around 1930, there were 214 tea houses in Deqing County. According to *An Investigation of County Conditions* in 1935, the registered tea houses in Deqing and Wukang were Jingshanlou and Gonghelou in the County Town and Gangxianglou and Taipinglou in Xinshi Town; Fangleyuan and Yongxinglou in Wukang Town, Wangyuelou and Shunfenglou in Shangbai Town. In addition, tea houses were also distributed in villages. In Zhongguan, a water village, there were 12 tea houses in the early years of the PRC. In Yangfen, a mountain village, there were 5 tea houses and 4 tea houses in the countryside in Sidu Town before the founding of the PRC. The policy of reform and opening to the outside world invigorated tea houses in Deqing County, expanding tea sales and flourishing economy.

There were tea houses in Towns of Dipu, Xiaofeng, Meixi, Xiaojin, Ancheng and Baishuiwan in mountainous Anji County, having high water quality.

Tea houses had been added. The registration data in 2005 showed that there were more than 600 tea houses in 3 counties and 2 districts under the jurisdiction of Huzhou.

Because of the long living history of tea house industry, people in Huzhou have cultivated many unique tea customs. Many old peasants used to get up at midnight and walked for a few miles to the nearest tea house to drink tea without being stopped by rainy days. Town people drank tea in tea houses every day and took a bath in public bathrooms at night. This is one of the tea customs of Huzhou people.

Before 1949, there was a custom of "drinking arguing tea". When two parties had a conflict or there was a family dispute, they went willingly to a tea house, drinking tea while arguing. The other tea customers took part in its discussion to judge the right and the wrong. How and when to correct the wrongdoings by the wrong party was decided on the spot. And the wrong party paid tea fees for all the tea customers to thank them.

Sometimes this custom could be taken advantage of as a bullying situation. For instance, in Shuanglin Town where it was hard for a non-native troupe to earn a living even if the troupe earned a living in many other towns and cities, there used to be performances and sometimes there were conflicts between the troupe and local audience in the evening. In this case, the two parties went to Fuyuan Tea House to drink arguing tea. Occasionally, in the process, there was fighting between the two parties. Usually, the tea customers helped the audience instead of the troupe without considering who was right and who was wrong. As a result, the troupe suffered losses. The troupe leader made an apology and paid the tea fees for all, and even unexpectedly, the leader had to pay for the previously broken tea kettles and cups. This was the image of "Drinking Arguing Tea Heroes" portrayed by Ding Cong, a famous cartoon painter. This custom totally disappeared after 1949.

There was the custom of Tea Doctor in Huzhou. It was originated in Xianfeng Period (1851-1861) of the Qing Dynasty after Taiping Heavenly Kingdom Uprising. At that time, tea houses were in a depression. Some bosses of tea houses discussed together to find a better solution to the depression. There were four popular actors of Suzhou *pingtan* (story-telling and ballad singing, a local art form of *quyi* in Suzhou). Their performances were welcome in Suzhou, Huzhou, and Shanghai. When there was such a performance in a tea house, the tea house was full of audience. So the combination of tea houses with Suzhou *pingtan* was a good solution to the depressing tea house industry. In the western part of Zhejiang Province, if conditions were permitted, the tea house boss tried to invite the performers of Suzhou *pingtan*. This form of story-telling and ballad singing was first practiced by Laoyihe Tea House in Huzhou, saving the depressing tea house industry owing to the fact that local customers welcomed this way of operating tea houses, enjoying tea and appreciating local tone of story-telling and ballad singing at the same time. So tea house bosses called these *quyi* performers tea doctors.

This custom of combining tea houses with story-telling has been kept.

In this tea hometown, its people are hospitable and simple especially in the countryside. When a stranger or a familiar one was in front of a house, its owner warmly invited him in and sent him a cup of tea for hospitality.

The custom of Three Cups of Tea would be practiced when the Dragon Boat Festival, the Mid-autumn Festival or the Spring Festival were approaching, or when there was a wedding party, a birth ceremony of a new-born baby, a birthday party for an adult, or a ceremony of new house building in the water region of Huzhou including Changxing in the west, Lianshi in the east, Deqing in the south and Taihu Lake in the north. The first cup of tea was sweet, symbolizing a sweet life in the year; the second one was salty, symbolizing good luck and colorful life; and the third one was pure, symbolizing the life philosophy of purity and simplicity.

Tea customs played an important part in wedding ceremonies in Huzhou. For example, when a boy and a girl were in love and their parents agreed with them, the boy's parents would send the girl's parents gifts or money, which was called *qiuhong* or *duan'anxinpan* meaning a good beginning or comfort. The girl's parents accepted the gifts or money, which was called *shoucha* meaning accepting tea. On the same day of accepting tea, the girl's parents would send gifts back to the boy's parents with a small bag of tea symbolizing water, and a small bag of rice symbolizing soil, meaning water in harmony with soil, hoping a smooth life of the bridegroom after marriage. On the marriage day, the bride went to the bridegroom house with his friends to take her, beating drums. When the team arrived at the bridegroom house, they practiced sit-tea custom, sitting around tables with candies, cigarettes and cakes, drinking "three cups of tea" from the bridegroom's assistants, and warmly welcomed by the bridegroom's father, the husband of her aunt and the brothers of her mother, full of joy and happiness.

In the past, for the majority of people, a marriage ceremony was held in the bride's and bridegroom's home. Since all guests and assistants would drink tea, the business of *chadan* (tea service) appeared. Tea service was a common practice for the reception of guests in the ceremony of marriage and funeral, operated by professionals. During the Republic of China, Yuan's tea service composed of Yuan Liankui, the master, and his apprentices, was very famous in Shuanglin Town.

Generally, tea service was set in the front hall or courtyard for the convenience of entertaining guests. The major tea tools were a tea furnace, a tea pot with a sling and tea cups, serving tea for about 100 people. When serving tea, a waiter held high a long-mouth copper tea pot with a sling and amazingly poured the tea into the cups in the distance without a drop of tea dropping, sometimes showing his stunt of pouring tea like a phoenix nodding three times. The guests' tea cups were white porcelain bowls made in Jingdezhen, the porcelain capital of China. The bowl with a cap and a saucer was beautiful and practical, showing the user's elegance. The male guests mostly drank new green tea, while the female guests preferred Xundou Tea made of a small amount of green tea and some baked cooked beans. Though it seemed quite simple to serve tea for guests, yet it was no easy to offer an on-time, nimble, satisfactory, safe and busy service in a crowded house with guests attending a marriage ceremony. Now, this service is still provided in the country side.

There was a custom of "Drinking Tea at Tudi (Land) Temple" in Nanxun Town. Before the Chinese People's War of Resistance against Japanese Aggression, the custom started near Tudi (Land) Temple in the town center after the Dragon Boat Festival in the fifth lunar month and ended in the ninth lunar month. A long shed was set up with one hundred tables and several hundred benches, creating a huge tea market. At the same time, restaurants would set their booths there to sell various fried dishes and snacks. There were puppet play and circus. On the fifth day of the ninth lunar month, high quality Peking opera shows were performed on invitation to thank Land God. Actually, these four months became the recreational months in Nanxun Town. 600 people at least and 1,000 people at most came here to drink tea every day. People, old or young, male or female, enjoyed themselves to their heart's contents. This was well portrayed by a stage couplet that read "Adults and children came here to kill their leisure daytime. They went back home to talk about it till midnight". The custom was put to an end after the outbreak of the War because the Temple was destroyed.

5.5 Overview of Historical Sites of Huzhou Tea Culture

5.5.1 Mt. Miaofeng Scenic Area

5.5.1.1 Sangui Pavilion

Sangui Pavilion was first built in October, 773, the 8th year of Dali Period in the Tang Dynasty by Yan Zhenqing, Prefect of Huzhou Prefecture, and was thus named by Lu Yu in the light of the three coincident *gui* on the date of its completion, namely, on the day of *guimao* of the month of *guihai* of the year of *guichou*. In the caption to his poem inscribed on "Sangui Pavilion in Mt. Zhushan", Yan Zhenqing said, "The pavilion was named by Lu Hongjian (Lu Yu)." Jiaoran also asserted that "the pavilion was named by Lu" in his caption to the poem in reply to "Stepping onto Sangui Pavilion of Miaoxi Temple" by Yan and Lu.

Located in the southeast of Miaoxi Temple, Sangui Pavilion stands high, providing a broad and far-looking view, and commands admiration of numerous poets through the ages. Take Yan Zhenqing's lines for example:

> Within the limited size of a dozen centiares,
> Its lofty peak kisses the clouds in the air.
> Graceful figure attached to the mountains high,
> It unfolds the ancient ferry with a strain of your eye.

Jiaoran's poem goes as follows:

Drooping down it wears the attire of water shiny like gold,
Looking up it kisses the sky with the mountain in its hold.
On the fresh landscape everything is new and bright,
Standing high and looking far the unique vision will itself unfold.
Leaning against the rock you forget the worldly cares,
Fondling the clouds you remember to pursue life's affairs.

During Kangxi Period (1662-1722) in the Qing Dynasty, Wu Qi, Magistrate of Huzhou, also sang high praise for its unique sight in his poem "A Sightseeing Tour to Mt. Zhushan":

Mountains and valleys have nothing different,
With interchange of cloud and fog highly frequent.
East and west are permeated with flowers of mist.
Yet the walls of the city are clearly evident.

Obviously, the second lines of the three poems illustrated that Sangui Pavilion was located in a high place. The last lines of the first and last poems suggest that the pavilion overlooked the ancient ferry and commanded a panoramic view of Huzhou.

Now the newly built pavilion is perched on the hummock east of Miaofeng Mountain with an altitude of 120 meters. Funded by tea lovers at home and abroad as well as relevant departments in the city of Huzhou, the pavilion was reconstructed in October, 1993, with an inscribed board of its name by Mr. Zhao Puchu (1907-2000, an outstanding calligrapher). To the north of the pavilion is a poem tablet carved with poems of Yan Zhenqing, Jiaoran and others. Approximately fifty meters in front of the pavilion is a stone tablet with the inscriptions of "Sangui Pavilion" by Mr. Sen Soshitsu, the master of Liqianjia of Japanese tea ceremony.

5.5.1.2 Lu Yu's Tomb

Lu Yu died at Qingtang Retreat in the late Zhenyuan Period (785-804) in the

Tang Dynasty and was said to be buried in Mt. Zhushan with Jiaoran's tomb on the other side of the valley. According to *Biography of Eminent Monks · Jiaoran* by Zanning in the Song Dynasty (960-1279), Jiaoran "died in the temple of the mountain". Meng Jiao's poem "Sending Lu Chang Back to Huzhou and Pondering on Jiaoran Pagoda and Lu Yu's Tomb" has the following lines:

Before the Temple heavy rain is captivating,
In the cold wind white flowers are shuddering.
Gone is the merrily crowded verse chorale,
Left now is the forever empty hall in the pavilion.
Chanting alone in the cold company of the jade,
My glaze wandered over the distant misty landscape.
Beside me are only the tombs solitary and cold,
With Jiaoran and Lu Yu sleeping on Zhushan Mountain.

The alleged temple refers to Miaoxi Temple. The above poems are evidence to prove that Lu Yu was buried in Zhushan Mountain.

The original tomb of Lu Yu has been lost beyond all recall and the details can never be verified. Sponsored by relevant departments in Huzhou and tea lovers both at home and abroad, the present tomb was reconstructed in 1995. It lies at the west side of Tongbao Dock in Miaofeng Mountain, facing the southeast and with a broad view, and standing against the green pines and verdant bamboos. In front of the tomb, there is Hongjian Bridge across the mountain creek with ascending stairs leading to the higher position on the other side of the creek, where the tombstone comes into view with the inscription of "Grave of Lu Yu, Educator and Minister of Education in the Tang Dynasty" by Mr. Wang Sunle, a librarian in Zhejiang Research Institute of Culture and History and an eminent calligrapher in Huzhou. Behind the tomb is a circular stone wall engraved with the full text of *The Classic of Tea*.

5.5.1.3 Jiaoran Pagoda

Monk Jiaoran, with the style name of Qingzhou and secular surname of Xie, was born in Changcheng (the present Changxing County) and the 10th generation descendant of the Song poet Xie Lingyun in the Southern Dynasty (420-589). With his birth in the early Kaiyuan Period (713-741) of the Tang Dynasty and his departure in the late Zhenyuan Period (785-804), he lived approximately between the years of 720 and 800. According to *Biography of Eminent Monks · Jiaoran* by Zanning in the Song Dynasty, "He was extraordinarily gifted in his childhood, with temperament and disposition well fit into the principles of Taoism. Having just started to take off the yoke and fetters of secular affairs, he needs only gradual and uplifting influence exerting on him." He once declared that "I would go into Zhushan Mountain and stay in company with pines and white clouds". In his middle age, he served as the presider of Miaoxi Temple and was hailed as "the Greatest Virtue Monk of Mt. Zhushan" by Yan Zhenqing.

Jiaoran was not only proficient at Buddhist philosophy, but also a master of poetry and tea study. There are 18 tea poems in his *Collection of Zhushan Mountain Poems*, *Poetic Structure* and others. He elaborates on tea matters in his poems relevant to the producing, picking, boiling and savoring as well as its function, contributing a lot to Lu Yu's compilation of *The Classic of Tea*. As the initiator of "Tea Ceremony", he makes the following remarks in his "Song of Tea: A Mockery of Cui the Prefectural Governor": "The first cup drowns my drowsiness" "A second refreshes my mind" "A third enables me to grasp the ultimate integrated truth of Taoism and tea" and "Who dares to claim to really and completely know the true meaning of tea? No other than the immortal Danqiu reaches this realm". Obviously, despite age difference, Jiaoran became an intimate friend of Lu Yu, the Tea Sage. He was also a brilliantly accomplished tea lover himself, worthy of the admiration of all the following generations.

After his death in the mountain temple, Jiaoran was buried in Zhushan Mountain in a ceremonial pagoda. The original pagoda is no longer recoverable or verifiable. The present one, rebuilt in 1996, is located in the east side of Tongbao Dock in Miaofeng Mountain, facing southwest, only a valley apart from Lu Yu's tomb. The two can be said "to live as intimate friends despite age difference and to die as permanent companions".

5.5.1.4 Zhaoyin Courtyard

Zhaoyin Courtyard was where travelling eminent monks, celebrities and influential pilgrims were entertained in Miaoxi Temple. It was also one of the places for the compilation and revision of the monumental reference book *The Complete Collection of Rhymes* by Yan Zhenqing, which was, according to historical records, started in the winter of 773, the 8th year of Dali Period of the Tang Dynasty and completed in the following spring. Among the 56 participant compilers, 19 were longtime dwellers including Lu Yu (the Saint of Tea), Fahai (the Buddhist monk from Jingling), Li E (the provincial censor), Chu Chong (an assistant educator in Imperial Academy), Tang Heng (the reviewer), Liu Cha (the commissioner of sacrifices from Qinghe County), Pan Shu (the deputy magistrate of Changcheng County), Pei Xun (the county security magistrate), Xiao Cun (the general secretary of Changshu County), Lu Shixiu (the security magistrate of Jiaxing County) and others. Another 27 people including Pei Yu, the official in charge of routine affairs, were "commuters", people who traveled regularly from home to the writing place. Besides, 10 other people including Liu Mao, the security magistrate of Wei County, dropped out midway before the book's completion due to various other urgent affairs to attend to. So in a sense the compilation of this monumental collection was so formidable a mission that it summoned up all the elites' involvement and devotion; on the other hand, it was self-evident that Zhaoyin Courtyard was a massive, grand and pompous building.

It is verified that Zhaoyin Courtyard was located in the east of Miaoxi Temple, to the north of the present Sangui Pavilion.

5.5.2 Mt. Guzhu Scenic Area

5.5.2.1 The Tribute Tea Courtyard

It was the first officially-run imperial tribute tea courtyard in our country. Starting from 770, the 5th year of Dali Period of the Tang Dynasty, it provided the tribute tea for both the imperial court and local governments for over 870 years on end.

Previously, Lu Yu had been to Mt. Guzhu on many occasions for field investigation of tea, often making evaluations of diverse tea breeds with Zhu Fang. They remarked that, among the varieties of choice tea, listed on top was Zisun Tea produced in Guzhu and they recommended it to the imperial court. Lu Yu also wrote *Tea Episodes of Guzhu Mountain*. Jiaoran managed a tea plantation in the above said mountain, as mentioned in his poem "To Pei Fangzhou while Visiting

Mt. Guzhu" that "I have fresh heavenly spring near Guzhu Mountain to which tea affairs over there are indispensable".

Ever since Zisun Tea became the tribute tea, there had been 28 Huzhou prefects coming in succession to Guzhu Mountain for the supervision of the tribute tea processing, whose efforts left behind such precious historical sites as cliffside carvings with their own inscriptions.

According to historical records, the Tribute Tea Courtyard used to lie behind the Hutou Rock. "Tea leaves from each town are baked in Guzhu under the supervision of the prefects and in the control and administration of the higher inspectors. At the very beginning there were over 30 thatched cottages there until a temple was built in 801, the 17th year of Zhenyuan Period of the Tang Dynasty, when Li Ci, Prefect of Huzhou Prefecture, found the courtyard too "narrow and shallow". The inscribed board by the name of Jixiang Temple (Auspicious Temple) in Wukang was moved here to be used as the name of the temple. The previous 30 cottages along the eastern corridor served as the tribute tea houses, thus achieving the harmonious unity of temple and tea courtyard.

The Tribute Tea Courtyard is lined on both sides with *dui* (also called *chujiu*, stone pestles and mortar used for mashing steamed tea leaves), equipped with more than a hundred baking ranges and staffed with over a thousand craftsmen and workers. "Guzhu spring water is diverted to the courtyard, indispensable to any single procedure like cooking, steaming or washing." In 845, the fifth year of Huichang Period of the Tang Dynasty, the temple was demolished due to the "Huichang Buddhist Hazard" (a national campaign to get rid of Buddhism). It was in 854, the eighth year of Dazhong Period of the Tang Dynasty that Zheng Yong, Prefect of Huzhou Prefecture, had it reconstructed under the imperial order. In 1337, the sixth year of Hongwu Period of the Ming Dynasty, Xiao Xun, the county chancellor, had it rehabilitated.

In the early years of the Tang Dynasty, the annual amount of the tribute tea was 500 strings (about 250 kg), all of which had to "be rushed to the palace batch by batch" before the Qingming Festival. This is the so-called "Urgent Tea". Li Ying in his poem "Song of Tribute Baking in the Tea Mountain" exclaims: "Ten days of horse riding to cover four thousand *li*, the tea must reach the imperial Qingming feast with no delay even for a while." In 781, the second year of

Jianzhong Period, the amount of the tribute tea was increased to 1,800 kg, and to 9,200 kg in 843, the third year of Huichang Period. In 838, the third year of Kaicheng Period, Prefect of Huzhou Prefecture submitted a memorial to the throne for additional three or five days of delivery deadline of "Urgent Tea" and his memorial was approved. Thereafter there was a slight relief in terms of the time limit.

By the Song Dynasty, the tribute tea had shifted its source of supply mainly to Jian'an in Fujian, only keeping a very small amount of Zisun Tea from Huzhou. For instance, in 978, the third year of Taiping Xingguo Period, only 50 kg of Zisun Tea was demanded as the tribute tea. The former Tribute Tea Courtyard was converted into a "Tea Grinding Courtyard" in the Yuan Dynasty, with 1,000 kg of powder tea and 45 kg of the sprout tea as tribute.

In 1375, the eighth year of Hongwu Period of the Ming Dynasty, Emperor Zhu Yuanzhang demanded only 1 kg of the sprout tribute tea. There was, however, no decrease in the local levy of tea. In 1405, the third year of Yongle Period, Guzhu Mountain still boasted of 1.8 *mu* of official tea fields with 14 tea-picking professionals.

By 1646, the third year of Shunzhi Period of the Qing Dynasty, because the local territorial despots hadn't been cracked down and the tea fields were mostly deserted, Liu Tianyun, Magistrate of Changxing, submitted a report of the status quo to the viceroy of Zhe-Min whose approval set free Zunsun Tea as the tribute tea.

In the spring of 1949, Jixiang Temple was burned down. In 1984 the people's government of Changxing County ranked the Tribute Tea Courtyard ruins among the county-level cultural relic protection units. In 1997, it was listed as the provincial relic protection unit and had a tablet erected on its site.

5.5.2.2 Jinsha Spring

Jinsha Spring, as another kind of the Tang Dynasty tribute, enjoyed equal popularity with Zisun Tea. At that time, the brewing of Zisun Tea was impossible without water from Jinsha Spring; therefore, the spring water was also given to the court together with the tribute of Zisun Tea. In the Tang Dynasty two silver bottles of the tribute water was paid and in the Song Dynasty, one bottle. The net weight of the silver bottle was 56 *liang*.

The geological exploration discovered that the geological rock composition from Guzhu Mountain all the way through to the town of Shuikou belongs to the granite strata and there is a large quantity of spring water stored 30 meters underneath the rock layer. So, the spring water does not come out only in one place; in fact, the water local people used for their daily life in these areas is spring water of the same quality. Jinsha (golden sand) Spring beneath Guzhu Mountain was thus named because of "the inflow of sands in the crystal water that are shimmering like Venus". There is endless outflow of water throughout the year, with more volume in the rain season and less in the dry season. In 1278, the 15th year of Zhiyuan Period of the Yuan Dynasty, Jinsha Spring unexpectedly "overflew overnight and irrigated fields of thousands of acres" and thus was bestowed the name of "Ruiying Spring" by Emperor Shizu (Kublai Khan).

Since the reform and opening-up in the late 1970s, through the identification of relevant departments, the Ministry of Geology and Mineral Resources issued a document in 1988 conforming that Jinsha Spring has "high quality mineral water containing strontium, metasilicic acid and radon". Due to its qualities like superb sensory performance, sweet taste, insulation from pollutants of chemicals and microbials, it offers a promising prospect for exploration and exploitation as drinking mineral water.

The present Jinsha Spring was redeveloped in 1984 by the People's Government of Changxing County.

5.5.2.3 Wanggui Pavilion

As recorded in the historical statistics, in order to meet the needs of gatherings of officials and men of letters, nearby the Tribute Tea Courtyard were erected several pavilions, terraces and towers in succession through the ages, the verifiable ones of which are Zhenliu Pavilion, Mugua Hall (Qinghui Pavilion), Qingfeng Tower, Xigong Pavilion, Piyun Pavilion, Yanggao Pavilion, Jinghui Pavilion, to name only a few. They have now all been destroyed and some are even left with no trace to be found. Among them was Wanggui (forgetting to go home) Pavilion, the most peaceful and graceful of all, first built in Zhenyuan Period of the Tang Dynasty. The present pavilion was constructed in 1984 in a new site with the budget provision of the People's Government of Changxing County. Here in this pavilion you can appreciate the green hills and water and enjoy the fragrant aroma of Zisun Tea to your heart's content.

In front of the pavilion is the tablet with the poem inscriptions of Mr. Zhuang Wanfang, a pioneer in contemporary tea study. The poem reads:

> Guzhu Mountain valley abounds with Zisun Tea,
> Its fragrance was already known in the Tang Dynasty.
> With a bowl of light tea conveying kindly intention,
> Merry mirth in poem jointing and chanting fills the pavilion.

5.5.2.4 Cliffside Carvings

At the end of supervising the tribute tea processing at Guzhu Mountain, the prefects of Huzhou Prefecture would have their tea poems and words about their supervision inscribed or carved on the cliffside in an attempt to hand them down to later generations. These carvings are now the historical witnesses of thousands of years of tea culture.

The cliffside carvings in Guzhu Mountain can be categorized into three groups covering nine sites, among which the most typical group is that in Waigang of Jinshan Mountain. With the inscriptions of Yuan Gao, Yu Di and Du Mu, prefects in different years, this group is hailed as "the highest hall" engravings. Yuan Gao's inscriptions were engraved in 784, the first year of Xingyuan Period and on the top of the stone, which reads, "Yuan Gao, the prefect in the Tang Dynasty, acting upon the imperial edict, came to the tea mountain for the supervision of the tribute tea and at the end of it stepped onto this highest hall to have a tea mountain poem engraved here on March the tenth of the first year of Xingyuan Period." There are 33 characters with one of them missing. Yu Di inscribed in the stone in 792, the eighth year of Zhenyuan Period, with the characters of "Under the imperial edict I came to Guzhu Tea Courtyard to supervise the tribute tea processing and at the end of it climbed onto this highest hall in Xigu Mountain". There should be a total of 68 characters with 6 of them missing. Du Mu's inscriptions made in 851, the fifth year of Dazhong Period contain such characters as "Du Mu from Fanchuan has completed his supervision job by the imperial edict". There are a total of 20 characters with 8 of them missing. Another group is

Baiyang Mountain Cliffside Carvings. Approved by Zhejiang Provincial People's Government in 1997, "Guzhu Mountain Cliffside Inscriptions (the Highest Hall Engravings)" was ranked among the provincial cultural relics protection units with a logo of cultural relic protection erected by Changxing County People's Government in 1999.

5.5.2.5 Xuwujie Wild Tea Fields

About 4 km west of Guzhu Village, there is a valley named Xuwujie (also called Ruwujie) with an altitude of over 300 meters and an incline of about 45 degrees. It is surrounded by exuberant groves of bamboos interspersed with broad leaf woods like sassafras and candlenut trees. Across the valley are pieces of land with weathered rocks formed through the piling up of mountain dirt that is scoured by floods. The land is covered with several patches of wild Zisun Tea trees, one of which is as tall as 2.5 meters and as old as over a hundred years. Some also grow on the banks of creeks or in the shade of trees or among the bamboo groves. By the way, the wild tea fields are also widely distributed in Xuhongjie and Huangkanling.

According to its composition analysis by the agricultural department, this kind of soil with weathered rocks contains 6.81% of organic substance, 0.347% of nitrogen, 0.0988% of phosphorus and rapidly-available potassium 128 ppm, pH value of 5. The relative humidity is more than 85%, with an annual rainfall of 1,500 mm and an accumulated temperature of 4,410℃ to 4,900℃ in terms of the yearly temperatures above 10℃. With the favorable climatic and geographical conditions, the soil is most suitable for the growth of tea plants and there is no need for artificial cultivation.

When he was making field investigation in the Guzhu Mountain area, what Lu Yu picked was exactly this kind of wild tea. He explained in *The Classic of Tea* in this way: "The top quality soil for tea growth is that with fully weathered rock, gravel soil is the second best and clay is the worst." "On the sunny slopes, purple bud leaves from those tree-lined plants are better than the green ones. Besides, slender bud leaves shaped like bamboo shoots are better than the shorter ones. In addition, those bud leaves still with curly blade are of higher quality than those already flatting out." These scientific assertions fully confirm to the reality of Guzhu Mountain and undoubtedly still have considerable realistic significance even to this very day.

5.5.3 Sites in the City of Huzhou

In the tenth year (775) of Dali Period, Lu Yu moved to Qingtang Retreat where he completed the revision of *The Classic of Tea* in the first year(780) of Jianzhong Period and where he died in the later years of Zhenyuan Period. In a word, the Retreat was his home in his middle and old age and thus was of paramount importance to him.

5.5.3.1 Qingtang Retreat

Qingtang Retreat lies near the Tiaoxi River outside the gate of Qingtang in the northwest of Huzhou. *Annals of Huzhou Prefecture* records that "Lu Yu's residence is located outside the Qingtang Gate" and quotes Jiaoran from his poem of "Delight in Yixing Prefect Quan Deyu's Arrival from Junshan and Accompanying Him to Lu Yu's Qingtang Retreat". In the Tang Dynasty there were nine gates in Huzhou, including the northwestern gate named Yingxi Gate or Qingtang Gate, commonly known as Qingtong Gate, outside of which was Qingtong Bridge. Jiaoran remarked in his poem "Failing to Meet Lu Yu in His New Abode": "He moved to the vicinity of the outskirt, country roads led to where mulberry woods were growing." This showed Qingtang Retreat was not far from the city, but was hidden in the mulberry trees and hemp plexus. What used to be the outside area of the old-time Qingtang Gate has now given way to Huzhou New District full of high-rise buildings and road network. Fortunately, there remains a patch of empty land on the bank of the Tiaoxi River, where a park is planned to be opened. Besides, Qingtang Retreat was reconstructed in 2003 in the northwestern direction of Qingtong Bridge with the approval of the municipal government and its provision of land and allocation of funds. This is another major move of Huzhou for the commemoration of Lu Yu and the promotion of tea culture.

Inside Qingtang Retreat are displayed the data of tea culture, and in the middle large frescoes presenting his life and major achievements in Huzhou. The frescoes are divided into five groups: "Brewing Tea while Writing *The Classic of Tea*" "The Completion of Sangui Pavilion" "Antiphon Poems in Wazun Pavilion" "Responsorial Poems on Xisai Mountain", and "Distinguished Guests from Afar". Besides, there are a few poem tablets embedded.

In front of Qingtang Retreat stands the statue of Lu Yu, 2.3 meters high, 1.65 meters wide, 1.42 meters deep, carved from a whole piece of granite with the gross weight of 13 tons. The statue weighs about 5 tons. It was elaborately engraved by Professor Da Liusheng from Zhejiang Academy of Fine Arts based on the design reviewed repeatedly and comprehensively by the experts and scholars at the invitation of Lu Yu Tea Culture Research Association of Huzhou. The image of Lu Yu is, as shown in the statue, fresh, dynamic and lifelike. This is the right manner and image of the host of Qingtang Retreat in people's imagination.

5.5.3.2 Sangzhu Pavilion

There stands a bamboo pavilion on the right mound behind Qingtang Retreat. It was named Sanzhu Pavilion with an inscribed board by Mr. Dai Meng. Inside it, a stone tablet was erected of "Record of Qingtang Retreat Reconstruction".

Sangzhu Pavilion is not a relic itself, but it has brought about two other sites into view and fame. One is the landscape of mulberry woods and hemp plexus around the original Qingtang Retreat, exactly for the sake of which Lu Yu named himself "Sangzhu Man". The other is the Sangzhu Garden that Wu Wenqi, Prefect of Huzhou, built in honor of Lu Yu in Wanli Period of the Ming Dynasty. As a native of Jingling himself, Wu Wenqi had the following account in his *Records of Sangzhu Garden*: "In the mid-Tang Dynasty, Lu Hongjian, also a native of Jingling, lived alone here far away from his hometown and thus found himself the style name of Sangzhu Man. He and I were born in the same neighborhood, traveled the same places and have the same interest in landscape and nature, so my garden is also his territory." Sangzhu Garden was originally located in front of Feiying Tower Courtyard situated to the northeast of the seat of the city government, but now is occupied with some modern structures and left with no trace to be found. By no means can any of the old-time mulberry woods, the ramie bushes or Sangzhu Garden surrounding Qingtang Retreat be recovered or reconstructed. Hence Sangzhu Pavilion was built for its commemoration.

5.5.3.3 Yunhai Tower

This was one of the places where Lu Yu participated in the compilation and revision of *The Complete Collection of Rhymes*. It was first constructed in 773, the eighth year of Dali Period of the east of the site where the present Renmin Park is in the center of Huzhou, namely, the east of the past government building inside the old Zicheng. The present tower was successively reconstructed in the periods of Tongzhi and Guangxu of the Qing Dynasty. In 1995 due to the need of urban construction, it was moved into Feiying Park exactly as what it was like.

5.5.4 Other Sites

5.5.4.1 Xisai Mountain

Xisai Mountain was first brought into fame by Zhang Zhihe's lines in his "Odes to a Fisherman":

> In front of Xisai Mountain are graceful egrets flying;
> In the peach blossomed stream are mandarin fish swimming.

It is said that this poem was the joint creation of Zhang Zhihe and others like Lu Yu, Yan Zhenqing, Xu Shiheng and Li Chengju by way of antiphon during their drinking party on Xisai Mountain. There was originally a total of 25 poems, with each of the five poets composing 5 poems. It's a pity that what remains of the antiphon are only those five by Zhang Zhihe and all the rest are no longer in existence.

Lu Yu made friendly associations with Zhang Zhihe on other occasions as well. At one time when Lu Yu was talking with Pei Xiu and Zhang Zhihe, Lu asked Zhang, "Who do you usually connect with?" Zhang replied, "With the universe serving as my chamber, the night moon sharing light with me, friends from all the four seas keeping my company, what's the use of other connections?" Another time Zhang Zhihe was painting for Yan Zhenqing on a white silk handkerchief, meanwhile Lu Yu invited a painting master to imitate the picture for permanent preservation.

Xisai Mountain is located 10 kilometers west of Huzhou, with Taohua Dock on the hillside and Fanyang Lake at its foot, which is connected with the Tiaoxi River. Now every spring, as always, peach blossoms would be in full bloom and mandarin fish would be the fattest. It is a good place for tourism.

5.5.4.2 Wazun Pavilion

Wazun Pavilion is located in Xianshan Mountain south of the city of Huzhou.

There are hollow stones shaped like wine goblets that can contain liquor. In Kaiyuan Period (713-741) of the Tang Dynasty, Li Shizhi, the attendant officer from Huzhou, used to climb up the mountain to have a drink with his friends. Later he had a pavilion built on the mountain and named it "Wazun". In 742, the first year of Tianbao Period, Li Shizhi was promoted to be the left prime minister, so it is also called "Left Prime Minister Li Stone Goblet".

According to historical records, in the spring of 774, the ninth year of Dali Period, a party of 29 people including Yan Zhenqing and Lu Yu came to Wazun Pavilion on Xianshan Mountain, creating "Antiphon Poems of Stepping on Xianshan Mountain and Appreciating the Stone Goblet of Left Prime Minister Li". Yan Zhenqing began the poem with the following lines:

For a mere drink Minister Li climbed the mountain
Where the goblet of a stone holds the attraction.

Lu Yu wrote the 19th line, which reads:

The deep pines attract leisured people into the wood,
With soft vines serving as grippers in danger to hold.

The number of the poets participating in this antiphon ranks No. 1 of its kind as recorded in the *Complete Collection of Tang Poems*.

Now Wazun, the hollow stone goblet, is still there, but the pavilion has gone forever. Other antiphon poems with Lu Yu's participation include "Farewell Antiphonal Poems to the Deputy Magistrate Pan in Water Hall" "Wind Antiphonal Poems with Geng Wei in the Water Pavilion" "Antiphonal Poems upon Listening to Cicadas in Youxi Pavilion" "A Tri-character Verse Conveying Delight in Meeting the Imperial Attendant Censor Huang and Appreciating the Moon on the South Tower" "Heptasyllabic Antiphonal Verse of Drunken Fancy" and "Leisure Time Antiphonal Poems to Lu Yu". However, the Water Hall, Water Pavilion, Youxi

Pavilion, South Tower and others have no trace to be verified.

5.5.4.3 Wenshan Mountain

The seventh chapter entitled "Records and Anecdotes" in *The Classic of Tea* quotes *The Records of Wuxing* by Shan Qianzhi, saying "Wenshan Mountain, 20 *li* west of Wucheng County in Huzhou, produces the tribute tea." Shan Qianzhi was a celebrity of the Song Dynasty of the Southern and Northern Dynasties. Based on the record in *The Book of Zhongxing of the Jin Dynasty* about Lu Na treating his guests with "tea and fruits" when he was the satrap of Wuxing, and according to the fact in *The Chorography of the Wu Kingdom · Biography of Wei Yao* about Sun Hao drinking tea rather than alcohol, it can be concluded that the latest time for Wenshan Mountain to produce the tribute tea should be in the Jin Dynasty (265-420) or even as early as the Three Kingdoms Period (220-265), more than 1,700 years ago.

Wenshan Mountain is situated on the side of Taihu Lake in the northwest of Huzhou, facing south and its main peak being 504.9 meters high. The southern side, with deep valleys, shady bamboo woods, moisture from water vaporizing, clouds drifting and mist enveloping, is perfect for the growth of tea plants. Since 1985 the famous Wenshan Imperial Tea has been reproduced, its base being expanded with each passing year.

5.5.4.4 Xiaoshan Temple

The seventh chapter entitled "Records and Anecdotes" in *The Classic of Tea* quotes *Sequel to Biographies of Eminent Monks* by a Buddhist with the following record: "There was a Buddhist monk Fayao in the Song Dynasty of the Southern Dynasty, whose secular surname was Yang, and whose natal place was Hedong. During Yongjia Period (307-312), he came down to the south of the Yangtze River and ran into Shen Taizhen who lived in Xiaoshan Temple in Wukang, to which Fayao was invited as a guest. At that time, Fayao was already advanced in age. In general, he drank a cup of tea instead of a meal. During Yongming Period (483-493), edicts came from the emperor instructing officials from Wuxing to ceremoniously escort Fayao to the capital city. That was the year when Fayao was already 79 years old." This indicates that tea is beneficial to human health.

The original Xiaoshan Temple was in present Yangkou Village, Wukang Town in Deqing County. It was destroyed many years ago. In 1996 the township government had a tablet erected marked with "the relic site of Xiaoshan Temple, Wukang".

5.5.4.5 Longquan Cove

Also known as Longquan Flatland or Baxue Cove, it is located in the town of Miaoxi in the southwest of Huzhou. Here the chains of mountains have graduated into flat land, wild tea covering the slopes ever since ancient times. Legend has it that Lu Yu used to come here to pick tea. This is now the maternal garden of famous tea in northern Zhejiang. In recent years, through both selecting from a local breeding and introducing from the outside a batch of quality varieties of tea breeds which are adequate to produce famous and superb tea, a production base of famous tea seeds is established, with tea gardens, covering an area of 700 *mu* of elite variety. In 1993, Lu Yu Tea Culture Research Association of Huzhou set up the monument of "Lu Yu Tea-Picking Place".

The above mentioned is the main historical events and relics of Huzhou tea culture, but is far from all. For instance, *The Autobiography of Lu Yu* talked about his "building a house on the side of the Tiaoxi River at the beginning of Shangyuan Period" and "often canoeing to and fro between the temples". This was his residence before he moved into Qingtang Retreat. As to the specific location, there is an assumption that it is outside the South Gate of Huzhou, and there are also those who believe that it was near Qili Pavilion outside the West Gate of Huzhou, neither of which has been confirmed. Take another example. The eighth chapter with the title of "Producing Regions" in *The Classic of Tea* lists tea "produced in the valley of the two counties of Anji and Wukang" but it is not clear which valleys it refers to. Besides, little is known about Lu Yu's activities in his old age. Some scholars maintain that Lu Yu, in his later years, once came to Jin'gai Mountain in the southern suburb of Huzhou for field investigation, which is most likely the case. In addition, Huzhou tea culture has a long history, but the question is: at what exact time did it start? There are now two mountains of Fengshan and Yushan in the town of Sanhe, Deqing County. Legend has it that they were once the native home country of a figure by name of Fangfeng, who became matchlessly strong after eating the local produce of baked cooked beans. If this were the case, the origin of Huzhou tea culture could be traced back to the time before the establishment of the Xia Dynasty in the 21st century BC. All this requires detailed investigations, exact verification and further exploration.

5.6 Books, Articles, Poems and Songs about Tea

Owing to the various wars during the Republic of China, though there were lots of tea drinkers, yet there were few tea books and articles passed down to the present. Tea books and articles began to appear and flourish after 1949 when the People's Republic of China was founded, especially after the establishment of Lu Yu Tea Culture Research Association of Huzhou in the early 1980s. Famous books and periodicals were *Selections of Tea Articles* edited by Lu Yu Tea Culture Research Association of Huzhou, *Collections of Tea Cultural Study* by Luo Jiaqing, *Huzhou Tea Poetry* edited by Zhu Nailiang. The 16 issues of *Lu Yu Tea Culture Research* edited by Lu Yu Tea Culture Research Association of Huzhou enjoyed a high reputation at home and abroad. Some of their articles had a great influence at home and abroad. They are "On Lu Yu" by Zhu Nailiang, "Examing Zhushan Mountain" by Zhang Baoming, "Analysis of Lu Yu's Thought and Character" by Kou Dan, "On the Plan of Developing Tourism Resources of Tea Culture in Huzhou" by Qian Pu, "On Tea and Non-tea" by Cai Yiping, "The Decline and Reinvigoration of Huzhou Tea Culture" by Gao Wanhu, "Examing Lu Lu's Traces in Huzhou" by Shao Yu, "On Lu Yu's Tea Poems Complexes in Huzhou" by Ding Kexing, "Ideas and Measures of Huzhou Tea Industrialization" by Lin Shengyou, "Harmony: The Soul of Chinese Tea Culture" by Zhang Xiting, and the articles written by Cai Quanbao about tea production and tea culture research in Deqing County. Some articles were read at international symposiums on tea culture or selected in the collections of tea studies. Some were published in academic periodicals such as *Agricultural Archaeology*, *Zhejiang Academic Journal*, *Tea Times*, *Journal of Huzhou Teachers College*, and *Huzhou Social Sciences*. Some were published on line and reprinted in other magazines.

The contemporary tea poems in Huzhou mainly sing praises of famous tea and springs in Huzhou. Sha Jin wrote "White Tea King":

White tea is famous in Anji, pure and white,
Its buds are yellow; its flavor delicate and subtle.
Tea bushes grow in high hills
Where tea leaves flourish.
The green hills are treasures
With tea plants absorb the natural essence.
If it were appraised by Lu Yu
White tea surely is No. 1 in China.

Dai Meng wrote"Mogan Yellow Bud Tea":

Springs flow in the hills and tea grows among bamboos,
Cloud and fog in Mogan breed Yellow Bud Tea.
Sword Pond shares the green tea hills,
And the tea cools thousands of people.

The above two poems describe the delicacy and subtleness of Anji White Tea and Mogan Yellow Bud Tea.

The famous spring in Huzhou is Jinsha Spring. There have been many poems praising it. Hu Aixuan, a modern poet, put it in one of his poems:

Zisun Tea and Jinsha Spring are praised for centuries.
The Spring, the treasure in the hills, is now used in people's home.

Lu Qian wrote "Jinsha Spring":

Ringing water flows from stones into a chilly pond.
Filtered by Jinsha Spring, the water is clean and clear.
Wood is used to fry crabs, and famous tea to intoxicate poets.
The famous tea is matched with the famous spring to best advantage.

Many other lines describing tea events can be subdivided into five categories. The first one is about tea chats and tea gatherings. For example, Ling Yi'an, in his poem "Hepanju Tea Chat in Huzhou", put it that "Cups of tea accompany spring. Intoxicated in tea flavor and poetic feeling, one is like a mortal." Xu Xuedong put it that "Before ascending Hualou, the tea house, my heart is intoxicated. After three cups of tea, I had no desire to go home." Du Shi'en put it that "The nature offers me the good tea. The oriental treasure nourishes the Chinese people." "Who understands tea men's intention? They taste new tea buds in the Qingming Festival."

The second category is to commemorate Lu Yu. Qian Shilin, attending the commemoration of the 1260th anniversary of Lu Yu's birth, wrote that "The Tiaoxi River still runs. The fame of *The Classic of Tea* is increasing. All the attendees are happy. And Sangui Pavilion appears new." When attending the commemoration of the 1270th anniversary of Lu Yu's birth, he wrote that "Qingtang Retreat was once again renovated. It was first built by Lu Yu. As years gone by, Mr. Lu passed away. His three volumes of *The Classic of Tea* found no rival." Zhong Weijin wrote in his poem of "Commemorating Lu Yu, the Tea Sage" that "At the Qingming Festival, 2003, people gathered to commemorate the Tea Sage sincerely and devoutly … Without songs and sacrifices but with tea and flowers, Lu Yu's style was spread".

The third category is about accompanied visits. In visiting Lu Yu's historical relics with foreign friends, Dong Shuduo wrote that "Sangui Pavilion welcomes domestic and foreign friends who praise Lu Yu and his glory in China." Qian Pu wrote in "Visiting Tea Hills with a Japanese Tea Group" that "Tea and poem entertain friends from afar. The tea flavor wafted across Japan."

The fourth category is about tea cultural tourism. A poem about visiting Sangzhu Pavilion by Ouyang Xun reads that "Ascending Sangzhu Pavilion, the air is fresh after rain ... Tea friends chat in the Pavilion, with happiness on their faces."

The fifth category is about tea customs in Huzhou. Zhu Nailiang wrote a poem about the custom of "Three Cups of Tea". It reads that "The sweet tea, salty tea with beans and pure tea with green buds are three delicious tea treasures. They are offered to guests and visitors. All praised them for they symbolize happiness, harmony and purity."

All these tea poems recorded tea activities in Huzhou in one way or another, being perfect and beautiful, sincere and happy.

Tea couplets can be regarded as another form of tea poems, usually seen in public places. In a pavilion for people to take a rest in Balidian Town of Huzhou, there was a tea couplet on the pillars carved in the late Qing Dynasty. It reads that "Everything will be nothing. Only a short moment here, strangers become familiar. Roads stretch at both directions. And after a cup of tea, everyone took his own road." The poem shows that at that time some people offered free tea to those who took a rest there. Tea couplets are more often seen in tea houses. Some are mounted and framed, hung in the tea house, and some are carved in wood pillars of the tea house. For example, "One is busy with fame, and busy with benefits; taking a rest while busy, one goes to a tea house for a cup. Working with brain is hard, and working with hands is hard; enjoying life despite hardship, one drinks a cup of tea." "Tea people come from different places. They gather in the tea house to become familiar."

The tea couplets in two tea houses in Zhicheng Town, Changxing County before 1937 are easy to understand but interesting and attracting. One of them, located in the north end of Zhouqiao Bridge in Zhicheng Town of Changxing County, was "Sanruoju Tea House". One of the couplets in it reads that "It is built near the Sanruo River. It looks at the remote Wufeng Hill." The other, located at the city gate of Dadongmen of Changxing County, was "Fangluge Tea House". One of the couplets in it reads that "Li Bai (a famous poet in the Tang Dynasty) went, shaking his head. Lu Tong (a poet in the Tang Dynasty) came, clapping his hands." In modern tea houses which are decorated in an ancient style, there are also some tea couplets.

There are also many tea proverbs in Huzhou. Some are about tea drinking.

For example, the proverb of "three meals and six teas" means a person should take three meals a day and drink tea six times or drink six bowls or cups of tea a day. This is in agreement with modern medical health; the proverb of "70% of tea, 80% of rice and 100% of liquor" means that when serving guests, it is okay to fill 70% of the tea cup for the convenience of adding water later and brewing the tea aroma, and it is proper to fill 80% of the rice bowl and 100% of the liquor cup, showing the host's or hostess' hospitality. The proverb of "drinking new tea but wine with years" means that it tastes better to drink fresh new tea and rice wine with years for its heavy aroma. Some are about tea picking. For instance, the proverb that "Tea leaves are time grass, and if picked three days earlier, they are treasure; if picked three days later, they are grass" talks about the importance of in-time picking of tea. Some are about tea growth. For instance, the proverb that "Tea grows thicker overnight and grows bones after the beginning of summer" means that tea grows too old to drink after the beginning of summer. The proverb that "Once defoliators appear in tea plants, tea peasants will take northwest wind" is about one pest which will deduce tea yield.

Most tea songs popular in tea areas are love songs. The following is "Tea Picking Song":

The first lunar month is the Spring Festival,
I bought a 12-*mu*-tea park without hesitation.
In the second lunar month, tea trees grew buds,
I begged you not to pick flowers when outside.
In the third lunar month, tea trees grew green,
I embroidered at the foot of them.
In the fourth lunar month, tea trees grew mature,
I was busy with housework and tea picking,
With silkworm breeding inside and wheat harvesting outside.

This song elaborates tea-picking girls' hard work and kindness.

A song entitled "Luojie Tea Song" is especially sung for Luojie Tea from Changxing County, describing hard work of Changxing people and the beautiful environment of Luojie Mountain.

There is an old temple, tea god and willow surrounding.

A spring runs in front of the temple and teas grow.

When spring comes, tea girls pick tea buds in Ruoling and Qifeng Mountains.

They are in front of the ridge and woodmen behind it.

The girls are singing and picking, with woodmen joining behind the ridge.

Their joyful singing scents Luojie Mountain.

They pick tea ridge after ridge, with baskets full of tea.

Losing their way owing to white clouds, they asked loudly for the woodmen's help.

Ridges and peaks surround tea parks,

And the streaming water in the gully sings better than orchestral music.

It rains in Qifeng Mountain while it is sunny in Dongshan Mountain,

The natural environment guarantees the high quality of Luojie Tea.

Another tea-picking song popular in Anji County sings:

Picking tea at Grain Rain, tea is budding.

It's hard to walk in a deep valley and stone ridge.

The husband picks more tea than his wife,

Picking another half kilogram of tea, we go back home.

Picking tea at the Dragon Boat Festival, tea is growing blue.

The husband and wife pass down the handkerchief among tea bushes.

The husband is singing folk songs and his wife joins in it,

Being affectionate to each other.

Picking tea at the Beginning of Autumn, tea is flourishing.
Tea blossoms scent the mountain.
After work, the husband sits in the shade of a tree,
And his wife nurses their baby.

This song sings for hardship of the couple picking tea in the mountains and their joyfulness in passing down the handkerchief, husband singing with wife joining, taking a rest and nursing their baby. Every section begins with a seasonal term: Grain Rain, the Dragon Boat Festival, and the Beginning of Autumn. Tea is picked three times there. Tea picked before and after the Qingming Festival is spring tea; tea picked before and after the Dragon Boat Festival is summer tea; tea picked before and after the Beginning of Autumn is autumn tea.

Popular in Santai Town, Deqing County, "Husband Is Starting Off to Pick Tea" vividly describes their romance, love and care between a newly married couple.

Husband: I am starting off to pick tea,
And my good wife stays at home.
Don't sit in front of the house without good excuse,
And comb less and don't wear flowers.
Give a smiling face to me when I come home.

Wife: My dear, my dear, don't be afraid of high sky and white clouds in the sky.
You love me and I love you. A fair heart is not afraid of burning fire.

Husband: My darling, my darling, don't say so too early.
Meeting realgar, a snake's teeth will become soft.
Seeing me, you are intoxicated.
Gold will melt in a furnace.

Wife: My dear, my dear, I will not let you pick tea,
For I am with child, and it will be born in August.
You will miss it day and night.
Husband: My darling, my darling, I have to go to a tea park.
Call it Chabao (tea treasure) if it is a baby boy.
Call it Chaying if it is a baby girl.
A baby boy or a baby girl, it is mine.

This folk song elaborates on the importance of tea in tea peasants' eyes. Their love, their future children and everything in their life are related to tea.

What's more, there is a tea tray song with 48 lines, beginning with the first lunar month until the twelfth lunar month. Anyway, it is not about tea events but about love. Unlike the general metaphor of flower for every month, a tea tray is used instead, showing the close connection of tea tray with everyday life of the people.

Some amount of tea folk songs are the reflections of tea peasants' wish, desire and hope for their good life, helping scholars to know about simple and ignorant souls and a sense of value of the tea peasants. In Chiwu, a mountainous town in Anji County, there is a custom of singing "Tea Luck" for a wedding ceremony in the countryside. Upon the end of the wedding ceremony, the bride and bridegroom are accompanied to their wedding room with friends and relatives singing "Tea Luck" and drinking lucky tea. For example,

When mentioning tea, we speak of tea,
There is a bud in tea.
The tea is from South Mountain,
The water is from Laolianhua in Longkou County.
The tea fire is from Bingding in the south,
And the fire wood is old pear branches from the north.
If an old man cooks the tea,
The longevity god wishes he could have a life of 880 years old.
If a sister-in-law cooks the tea,

She is expected to give birth to a fat baby.
If a young scholar cooks the tea,
He is expected to be No. 1 in the enrollment list of Imperial Examination.
If an iron master cooks the tea,
He is expected to be rich by making hoes.
If a cow boy cooks the tea,
He is expected to come back home safely in the evening.

"Ode to Tea Buds" was popular in Jiapu Town in Changxing County. Some extracts are as follows:

Tea-brewing people have a life of 1,000 years;
Tea-drinking people have a long life.
The flowers of white tea are golden yellow;
Tea and tea folk songs keep devils out of their life.
Drinking tea helps the old a longer life;
Drinking tea keeps the young smarter.
Every year silkworm has a good harvest;
Silkworm cocoons are like silver hills and their silk is like clouds.
If the fields are covered with green tea trees;
The family has a good harvest of grains.

These two folk songs about tea, though written in different towns, had similar understanding about tea. Both relate tea to longevity. This is no coincidence because the practices of generations of people with tea confirm the intrinsic relation between longevity and tea. In addition, Huzhou people know that drinking tea is good to child birth, beneficial to intelligence of children, and helpful to a good fortune and safety.

The hardship of tea production is also reflected in the folk songs. For example,

Peasants pick tea at daytime and worry at night.

If it rains three consecutive days in a tea-picking season, tears fill peoples' eyes.

In a tea-picking season, peasants keep their back bent and eyes blooded,

And their feet are hurt and bloody.

Doing one season tea jobs results in a disease.

Doing one year tea jobs costs half a life.

It accords with the saying that every single grain in the dining plate means hardship. So, we should treasure and save tea.

5.7 Non-governmental Tea Organizations

The major non-governmental tea organizations are Lu Yu Tea Culture Research Association of Huzhou, Tea Horticulture Institute of Huzhou, and Huzhou Tea Industry Association.

5.7.1 Lu Yu Tea Culture Research Association of Huzhou

Set up on October 24, 1990, Lu Yu Tea Culture Research Association of Huzhou is a non-profit and non-governmental academic organization. It aims to study Lu Yu, conduct academic exchange about the study of Lu Yu, promote historical culture of tea study, and keep in touch with people in the field of tea study and culture both at home and abroad for further friendship and the development of tea science and economy. Great progress has been made in its academic study, tea cultural exchange and promotion of famous tea development and tea economy.

5.7.1.1 Academic Research

As a non-governmental academic association, it focused its attention on academic research. By the end of 2005, the Association had edited and published

16 issues of *Lu Yu Tea Culture Research*, with one issue every year. About 900 articles with over 3 million words were published, and over half of them were written by its members. The papers contributed by tea experts at home and abroad increased the journal's universality and some papers by its members were read at international symposiums on tea culture or collected in the paper collections of these symposiums. Some were published in *Agricultural Archaeology*, *Zhejiang Academic Journal*, *Tea Times*, *Journal of Huzhou Teachers College*, and *Huzhou Social Sciences.*

5.7.1.2 Investigating and Examining the Historical Traits of Lu Yu's Life in Huzhou and Tea Culture

As the adopted home of Lu Yu, the Tea Sage, and an origin of tea culture in China, Huzhou has long been noticed in the field of tea study. So it is of great importance to track down historical traits of Lu Yu's life in Huzhou and tea culture. The Association has attached great importance to this examination since its establishment. For more than ten years, it has tracked down Miaoxi Temple in Zhushan and Qingtang Retreat where Lu Yu once lived and stayed, and the places where Lu Yu once was for his observation of tea by collective examination and individual research, by consulting literature and an on-the-spot investigation, and by repetitious analysis and demonstration. As a result, a batch of valuable articles were written and made public. The debates were illustrated by convincing facts. Based on these investigations and examinations, approved by government leaders and sponsored by relevant government departments, in 1993, Sangui Pavilion, Lu Yu's Tomb, Jiaoran Tower, and Muyu Archway at Miaofeng Mountain, the landmark of picking tea by Lu Yu at Longquanwu, and Qingtang Retreat at the west end of Qingtong Bridge were reconstructed. At the same time, in Changxing County, great efforts were made to build Zisun Tea Park, a stone tablet for the historical site of Tribute Tea Academy in the Tang Dynasty at Guzhu Mountain was again erected, the cliffside inscriptions of tea culture in the Tang Dynasty were primarily sorted out, and funds were allocated for the reconstruction of Tribute Tea Academy. In Deqing County, a tablet for the historical site of Xiaoshan Temple in Wukang, a tablet for the origin of baked bean tea and a tablet for tea customs in Zhongguan were built respectively in Luoshe, Sanhe and Zhongguan. In Anji County, Tea Sage Monument was built before Lingfeng Temple. In recent years, the two historical sites of tea culture at Guzhu Mountain and Miaofeng Mountain have attracted many tea culture fans at home and abroad. They would play a greater role in tea cultural activities in future.

5.7.1.3 Academic Exchanges of Tea Culture with Domestic and Foreign Counterparts

With the rise of tea culture around China and other parts of the world, tea cultural exchanges increase. The Association has established contacts with many tea societies and experts at home and abroad, and attended lots of tea activities. Its members participated in the First, Second, and Third International Symposium on Tea Culture held in Hangzhou, Zhejiang Province, Changde, Hunan Province and Kunming, Yunnan Province in 1990, 1992 and 1994 respectively. As one of the initiators of the Symposium on Chinese Tea Culture held in Wuyi Mountain, Fujian Province in 1991, the Association made a keynote speech. Noticeably, it co-hosted the 1260th anniversary of Lu Yu's birth with Zhejiang International Cultural Exchange Society in Huzhou in November, 1993, attended by 300 scholars and experts of 7 delegates from Japan and the Republic of Korea, from Hong Kong and other 9 provinces of China. Six delegates had performances of tea arts. Exhibitions of famous tea, tea sets, and paintings were held. On the exhibitions, there were 31 varieties of high quality famous tea of Huzhou, 1,100 pieces of art crafts made from Yixing clay. Some participants visited Miaofeng Mountain for the unveiling ceremony of the rebuilt Sangui Pavilion. Some visited the tea area in Guzhu Mountain, Changxing County. This activity was listed "Top 10 News in Huzhou in 1993" for its active role in enhancing the popularity of Huzhou and promoting economic construction and cultural exchange. Commemoration activities were organized for the 1200th anniversary of Lu Yu's *The Classic of Tea* and the 1270th anniversary of Lu Yu's birth in 2000 and 2003 respectively. With the Association's support, a symposium on *An Exposition on Tea* by Zhao Ji and Anji White Tea was successfully held in Anji in 2005 to promote White Tea development. The "Zhongda Cup" Qingtang Tea Essay Competition co-sponsored by the Association and *Huzhou Evening News* got good response. Tea industry and tea economy were vigorously promoted. The Association made tea culture popularization combined with tea market expansion, tea industry upgrading, and developing tea tourism and tea economy. This promoted the development of Huzhou economy and society. In April, 2005, it held Qingtang Tea Symposium attended by 500 domestic and foreign experts of tea cultures, tea scholars and others, making Huzhou known to more people.

From 1990 to 2005, the Association received 50 batches of foreign friends with 580 persons and 24 batches of domestic friends with 300 persons. Up to now, it kept frequent contact with 20 foreign tea culture groups and tea experts and scholars from 30 provinces and municipalities, and most of them are world-renowned in the tea field.

5.7.1.4 Helping to Develop Famous Teas and Tea Sets

The Association tried its efforts to promote tea economy in Huzhou. Its journal published many articles about tea economy and tea sets, introducing famous tea, tea-trading companies, tea and tea set factories in Huzhou. It held special lectures on tea culture in Huzhou TV Station, and tea package exhibition together with Huzhou Association of Package. It also co-held famous tea appraisal with agricultural departments. Based on investigations, it put forward tea tourism suggestions to relevant government departments in different forms and from various channels. It introduced foreign businessmen and businessmen from other provinces to tea and tea set factories in Huzhou. Many of its members were engaged in producing and developing famous tea and tea sets, and participated in small and medium tea competitions held by counties and towns. All these activities, embodying the spirit of serving economic construction, directly or indirectly promoted the development of tea economy and tea sets made from Yixing clay.

5.7.1.5 Establishing and Improving Its Operation Systems

After 15 years' development, the Association was expanded from 74 to 212 members at the end of 2005. In order to keep its normal operation, under the leadership of its council, it set up an administrative office, a liaison office, an academic office and an editorial board of its journal. According to its operation systems, important decisions and large-scale activities were made collectively by its standing council. The important daily work was also decided or solved by in-time work meeting or director meeting. Owing to the eagerness of its members and directors to serve tea culture, their responsibility, hard work and dedication, the Association did an effective and efficient job and won good social fame, awarded model association by Huzhou Federation of Social Sciences for several times.

5.7.2 Tea Horticulture Institute of Huzhou

Set up in March, 1984, Tea Horticulture Institute of Huzhou is an academic society of tea (fruit, vegetable) technicians.

It aims to develop, innovate and popularize tea horticulture technologies by working with tea technicians in Huzhou, focusing on economic construction, adhering to the idea that technology is the first productive force, and carrying out the strategy of sustainable and scientific development so as to train more tea technicians, promote the combination of tea technology and culture with economic development, accelerate industrialization of tea production, and contribute to the modernization of tea production. Since its establishment, the Institute has organized and participated in many important tea activities.

5.7.2.1 Tea Academic Exchange and Technological Training

The Institute held various forms of tea academic activities to exchange information, improve understandings about tea and guide scientific decisions on tea production.

On March 8, 1984, the Institute held a tea lecture about information of tea production and sale in Huzhou. This lecture played an important role in the study of tea information, in broadening envision on the tea market, and in developing multiple channels of tea marketing.

In early August, 1984, the Institute played host to a tea delegation of 13 Japanese from the City of Shimadashi with Professor Nunome Chofu and Mr. Matsushita Satoru as the leaders. It had dialogues and academic exchanges with Japanese scholars on *The Classic of Tea* by Lu Yu, the study of tea culture and tea technologies. It also accompanied them to visit such historical sites as Tribute Tea Academy at Guzhu Mountain, Wanggui Pavilion, Jinsha Spring and the production base of Zisun Tea. This was the first time for tea scholars and technicians of the Institute to exchange academic information about tea with foreign counterparts since 1949. After that, more academic exchanges were conducted with other tea scholar groups from Tianmen City, Hubei Province, and Japan.

In May, 1986, members of the Institute took part in the First Symposium on Lu Yu held in the City of Tianmen, Hubei Province. At the Symposium, Lin Shengyou read his paper entitled "Lu Yu in Huzhou", introducing for the first time

Lu Yu's historical traits, famous tea and tea culture in Huzhou to scholars about Lu Yu study and tea culture, and introducing for the first time the idea that Huzhou was Lu's adopted home, which inspired the participants' great interest. Kou Dan read his paper "On Zisun Tea and Jinsha Spring". These two papers were both published in paper collection of the Symposium. At the Symposium, Lin Shengyou, Kou Dan, Ding Kexing and Cai Quanbao were elected members of the First Lu Yu Study Society.

In January, 1987, the Institute and Huzhou Agricultural Bureau co-held a discussion on Lu Yu Study. Twenty-two people attended the discussion. They were experts, scholars and celebrities of literature, history, journalism, archeology, economy and tea science and technology from 18 units including Huzhou Committee of the Chinese People's Political Consultative Conference (CPPCC), Huzhou Agricultural Bureau, Tea Horticulture Institute of Huzhou, Lu Yu Tea House, etc. All the participants unanimously agreed that it is of realistic and strategic significance to study Lu Yu, a historic celebrity, for the increasing popularity of Huzhou in China and other parts of the world and for Huzhou's economic development. They discussed the advantages, its methods and organizational forms of studying Lu Yu in Huzhou. The participants concluded that Lu Yu and his *The Classic of Tea* enjoy high prestige in tea study in China and other parts of the world; that Huzhou is Lu Yu's adopted home, and Lu Yu contributed a lot to Huzhou and its people; and that it is realistic, necessary and urgent to study Lu Yu in Huzhou. They put forward some suggestions on better study of Lu Yu.

Supported by relevant government departments and tea production units, the Institute organized 40 tea-training classes attended by 3,686 trainees from 2003 to 2005. Professors and experts were invited to be the trainers from Chinese Academy of Agricultural Sciences, Zhejiang University and Zhejiang Provincial Department of Agriculture. Its members and the leaders of the demonstrative programs were sent to attend International Symposium on Sulfur and Potassic Fertilizers, Zhejiang Provincial Training Class of Pollution-free Tea Production and Its Trade, and of Tea Factories Optimization.

5.7.2.2 Participation in Science Popularization Week and Academic Activities

Huzhou Science Association organized the science popularization week and academic activities every year. The Institute actively organized its members to popularize science and take part in scientific consultation, and its member units and tea companies to popularize scientific and tea knowledge by way of exhibition board and product exhibition, making themselves and their products known to people. The Institute and its members tried every effort to make known to citizens pollution-free and organic agriculture and exhibit pollution-free produce so as to familiarize peasants with the advantages and features of pollution-free produce, with new technology and varieties of pollution-free agricultural production, promoting the development of pollution-free tea industry in Huzhou.

5.7.2.3 Active Participation in Technological Guidance of Tea Production and Industrial Development

The Institute helped relevant government departments and enterprises formulate procedures of production technologies and product standards. It introduced and popularized new improved varieties of tea and practical technologies. It conducted technological cooperation with tea factories and tea farms affiliated to its unit members. Making best of its technological advantages, it conducted technological consultation and guidance to help companies overcome the lack of technology and train technicians. All these efforts improved the level of tea production and accelerated the growth of tea industry.

5.7.2.4 Organizing Exhibitions of Tea Products and Tea Awards

The Institute and relevant government departments co-held appraisal of famous tea in June, 1984 and 1986, selecting a batch of prefectural level famous tea, high-quality flat teas, hand-made tea and machine-made tea. This played an active role in exchanging technologies and experiences of famous tea production, improving tea techniques and quality, and developing famous and high quality tea.

In the spring of 2003, co-working with Huzhou Agriculture Department and Lu Yu Tea Culture Research Association of Huzhou, the Institute took part in Lu Yu Cup Appraisal and Exhibition of Famous Tea. After tea experts' appraisal, Anji White Tea with the brand of Dashanwu and other four famous tea were awarded gold medal, 10 kinds of famous tea including Anji White Tea with the brand of Yuyu silver medal and 23 kinds of famous tea including Zisun Tea from Fangwu Tea Factory bronze medal. The medals and certificates were given by China Society of International Tea Culture Study and Huzhou People's Government. Co-working with agricultural sectors, the Institute organized tea businesses from the four counties and districts of Anji, Changxing, Deqing and Wuxing to participate in "2002 China Exquisite Famous Tea Expo" in Hangzhou and "2003 Shanghai International Tea Culture Festival & China Exquisite Famous Tea Expo". 30 tea businesses from the four counties and districts were organized to take part in "2004 Shanghai International Tea Culture Festival & China Exquisite Famous Tea Expo" on 12-15, April. In the exhibition and appraisal of famous tea, 20 kinds of tea were awarded. Among them, Anji White Tea with the brand of Dashanwu and other 7 kinds of tea won the gold medal. Mogan Yellow Bud from Deqing and other 6 won silver medal. Another 5 kinds of famous tea won high-quality tea medal. From late April to early May, 2004, it organized more than 10 tea businesses from counties of Anji and Deqing to attend "Ningbo China International Famous Tea Expo" co-held by China Tea Association and Ningbo People's Government. Several of them won "Zhonglü Cup" gold and silver medal teas.

In the appraisal of top 10 famous tea in Zhejiang Province in 2004, White Tea from Anji and Zisun Tea from Changxing were in the final contest. And White Tea was awarded one of top 10 famous tea in Zhejiang. On August 12-16, 2004, sponsored by Zhejiang Provincial Department of Agriculture, exhibition and marketing of Zhejiang famous tea were held in Hong Kong as part of the International Food Expo. Six members of the Institute, such as Lin Shengyou and Hu Zhengjian, attended the exhibition together with local leaders from tea production areas. White Tea from Anji was the favorite of Hong Kong citizens and foreign businessmen.

In addition, famous tea from different companies in Huzhou were seen in Zhejiang Agriculture Expo, (Wenzhou) China Produce Expo, (Xiamen) China, Japan & the Republic of Korea International Tea Culture Exchange & Contest of Tea King, and some other tea exhibitions held in Beijing, Ji'nan and Guangzhou. Some of them were awarded gold and silver medals. These expos and exhibitions showed the charm of famous tea from Huzhou, increased their popularity and raised their competitiveness, helping achieve the purposes of popularizing famous tea and tea brands from Huzhou and promoting the development of Huzhou's tea industry.

5.7.2.5 Tea Investigation and Study

In 1984, the Institute organized an investigation group about Yuchuan Tea from Wenshan, the earliest recorded tribute tea in Zhejiang and with a history of over 1,700 years. The group conducted a field study and consulted lots of literature. The investigation report "Examination of Wenshan and Wenshan Tea" provides historical and scientific evidences for reproduction and development of the tea.

The leaders of the Institute together with its members and tea technicians investigated tea production areas and tea farms, completing several investigation reports. The reports are "The Investigation of the Status Quo of Anji White Tea and Its Development Measures" "An Investigation of Dilemma of Developing Organic Tea in Nanwushan Tea Farms in Anji" and "An Investigation of Developing Bases for Zisun Tea in Changxing".

5.7.2.6 Carrying out Key Tea Technology Research and Popularization Programs

The Institute implemented two programs sponsored by Huzhou Association of Science & Technology. They were "Demonstration and Popularization of Pollution-free and Organic Tea Production Technology" and "Research and Popularization of Improving Technologies for Famous and High-quality Tea". Their completion helped achieve 20% yield increase of famous tea, double the yield value of famous tea, increase quickly the base areas, yield and economic benefits of pollution-free and organic tea, and accelerate the fast development of pollution-free and organic tea in Huzhou. These two programs won the third prize of their kind.

5.7.3 Huzhou Tea Industry Association

Set up on June 25, 2003, Huzhou Tea Industry Association was a non-profit organization mainly attended by businesses of tea production, processing and circulation, and tea households based on the principle of willingness, autonomy and mutual benefit.

This organization was operated in a way of self-operation and service and democratic management. Its members have equal rights, and share benefits and risks.

It bridges government and tea businesses, tea market and tea production bases, playing an active role in strengthening tea industry services, self-discipline and management, and contributing a lot to the industrialization and modernization of tea industry. Its main activities are as follows.

5.7.3.1 Investigation and Study

On December, 17-19, 2003, the Association organized an 11-person group to study tea industry in the south of Zhejiang Province. The group visited clonal tea bases, famous teas bases, Zhe'nan Tea Market, Songyang Fresh Leaves Market in Lishui City and Songyang County. They also visited famous tea bases developed in large fields in agricultural restructuring in towns of Xinxing and Zaitang, Songyang County, and there they visited the modern tea production and processing park solely invested by Japanese businessmen. Besides, they had a talk with Lishui Agricultural Department, Songyang Agricultural Department, Xinxing Town Government, and the leaders, tea traders and technicians of Zhe'nan Tea Market.

The leaders of the Association, its members and tea technicians did field investigations in tea farms and tea factories in Huzhou. They completed a special investigation report on tea-species in Huzhou, and a study report on tea-processing environment and raw tea-processing factories, providing reliable data for the government to guide tea production and formulate the 11th Five Year Plan for Tea Development. In May, 2003, it organized its council members, other members, tea technicians and heads of tea factories to study tea marketing in North China in Ji'nan, the capital of Shandong Province, trying to expand the market for Huzhou tea.

5.7.3.2 Tea Technological Exchanges and Training

Supported by relevant tea departments and tea factories, the Association held forty tea training classes with 3,686 trainees. Part of the lecturers was chosen from its members. Some were experts and professors invited from Chinese Academy of Agricultural Sciences, Zhejiang University, Zhejiang Provincial Department of Agriculture. Some of its members and the heads of demonstration tea factories were sent for attending Zhejiang Provincial Training for Upgrading Tea Factories. On March 19, 2004, it co-held with Huzhou Bureau of Agriculture an agricultural technology lecture given by Chen Zongmao, a member of Chinese Academy of Agricultural Sciences, attended by over 100 persons including heads and officers of County Agricultural Bureau, tea technicians and some members from County or City Tea Institutes.

5.7.3.3 Cultivating Tea Brands and Expanding Tea Markets

On May 20-23, 2005, it organized a group of 30 persons to attend "2005 China Competitive Famous Tea Expo" in Ji'nan, the capital city of Shandong Province. The members were heads of 12 tea factories, 5 tea cooperatives, tea stations and County or City Agricultural Bureau, and tea technicians. The group attended famous tea exhibition and appraisal, tea information exchange and forum on tea economy. It also took part in tea performances and famous tea recommendations held in Spring Square in Ji'nan. Besides, it visited Ji'nan Tea Trading Market, the largest one in North China. In this exhibition and appraisal were such famous tea as Anji White Tea from Yangjiashan and Lingfengshan tea farms; Zisun Tea from Changxing; Mogan Yellow Bud Tea, Mogan Jianya (Sword-like Bud) Tea, Mogan Qinglong Tea from Deqing County; Hutan White Tea, Gucheng Qiqiang Tea from Wuxing District. Anji White Tea, Mogan Yellow Bud Tea, Mogan Qinglong Tea, Zisun Tea were awarded gold medal.

5.7.3.4 Technological Cooperation and Support

Making the best of its technological advantages, the Association conducted technological consultation and guided tea factories and tea farms to solve their technological problems and train their own technicians.

Appendix

1. A Chronology of Lu Yu's Life

Lu Yu, whose birth has always been a mystery and involved a variety of legends, is said to be a foundling. Naturally, no one knows for certain about his family background. In the seventh chapter of his monograph of *The Classic of Tea*, he ranks people like Lu Na among his remote surnamed ancestors, following the convention of the powerful and influential family. Lu Na was once the satrap in Huzhou and Lu Chu, the son of his brother, was the satrap in Kuaiji.

According to the local chronicles, his life is richly legendary. An account is given in *Annals of Tianmen County* compiled in the period of Daoguang (1821-1850) that "A monk got up in the early morning and heard the honking of a flock of wild geese beside the lake trying to cover a baby with their wings. He adopted the infant." The "early-rising monk" anecdote is about Zhiji, the Buddhist monk from the previously so-called Longgai Temple (later renamed Xita Temple). He was walking in early morning at the bank of the West Lake in Jingling (Hubei) when he heard the loud noise of a flock of geese. Following the sound and coming closer, he spotted three geese protecting a baby with their wings and he carried him to his temple for adoption. Later, there appeared a succession of scenic spots and historical sites in Tianmen County, such as Yanjiao Pass, Guyan Bridge and others, all of which were analogies drawn on the lines of this legend.

Soon afterwards, to choose a proper name for the baby, Monk Zhiji resorted to *The Book of Changes* and was offered a divination of *jian* from the *Jian*

Hexagram, which reads: "Flocks of geese landing from the air, their feathers can be used as ritual gifts." Then in the light of the hexagrams, he found the baby a surname of Lu, a given name of Yu, and a style name of Hongjian.

Lu Yu was born in 733 AD (the 21st year of Kaiyuan Period of the Tang Dynasty), the year of *guiyou* in the Sexagenary cycle (as named by the Heavenly Stems and Earthly Branches). So in his hometown there is a kind of belief that Lu Yu is a baby born in the year of rooster. As is said by an elderly man with the monastic name of Xijiang Busou, who lived in the early years of the Republic of China (1912-1949), Lu Yu was probably born on August 15, the mid-autumn night. For lack of proof, this statement is generally not cited as kind of established evidence. In terms of Lu Yu's life experiences, there are truly some facts whose validity can be adequately confirmed by the existing detailed historical records available, starting from his looking after the cattle to trying his hands on writing articles at around 9 years old. From his early childhood, he had always been studious, boiling tea for Zhiji and doing the house chores. Because of his refusal to be converted to Buddhism, he suffered from servitude and torture. At eleven, he ran out of the temple and began to learn acting. He was humorous and flexible, especially popular for his role-playing of a "pseudo officer". In the meanwhile, he was also showing himself as an oncoming playwright. As such, he would manifest to the outside world his remarkably brilliant talents once stepping out of the confinement of his present condition.

In 746, Li Qiwu, a former prefect of Henan, was exiled to Jingling where he was to serve as the satrap. The host county magistrate planned to give the newly appointed satrap a performance of welcome and as luck would have it, the appointed acting troupe was the one that Lu Yu was working for. Greatly impressed by Lu's performance in the play of "Army Chief Inspector", the prefect personally summoned him, presented him with books and introduced him to Master Zou for schooling in Huomen Mountain, northwest of Tianmen. In his spare time after reading, Lu Yu also boiled tea for Master Zou.

In 753, at 20, he made the acquaintance of Cui Guofu, a former official in the Ministry of Rites, who was now demoted as Sima of Jingling. The two of them traveled together for three years.

In the spring of 754, Lu Yu bid farewell to Cui, setting out on a tour of

Yiyang in Henan and the Gorge of Sichuan. It was on the eve of the Qingming Festival that he got to Yiyang Prefecture (now Xinyang in Henan). Later in Bashan, he picked and tasted "Badong authentic fragrant tea". That was the time when he got a chance to acquire some knowledge about tea production in the Shu areas like Pengzhou, Mianzhou, Shuzhou, Yazhou, Luzhou, Hanzhou, and Meizhou. Then, he moved on for another detour to Yichang where he savored the taste of Xiazhou Tea and Hamaquan Spring Water.

In the summer of 755, Lu Yu returned to Jingling and settled down at the village of Donggang at the side of Songshi Lake near the Qingtan Station, an ancient post 60 *li* away from the county seat. There he sorted out what data he had obtained during his outings, conducted elaborate studies and got started on the compiling of his great monograph of tea.

In 756, due to the An-Shi Disturbances, Guanzhong (the Central Shaanxi Plain) refugees kept surging down. Faced with a half-broken homeland and numerous homeless people, he accomplished "Poem of Four Sorrows" to vent out his indignation. After he got across the Yangtze River with the Qin people, Lu Yu traversed across the middle and lower reaches of the Yangtze river and up to all the areas around the Huaihe River, investigating and accumulating a vast amount of information on tea picking and processing.

In 760 he arrived in Huzhou. He first stayed in Miaoxi Temple in Zhushan Mountain. There he acquainted Monk Jiaoran and established a "cross-age friendship between monk and layman". Soon afterwards, he moved to the Tiaoxi thatched cottage, fully concentrated on his writing. Meanwhile, he made acquaintance of such prominent monks and scholars as Lingche, Li Ye (also named Li Jilan), Meng Jiao, Zhang Zhihe, and Liu Changqing.

In 765, based on the data derived from his field investigation of the 32 prefectures and through years of strenuous study and detailed research, Lu Yu finished the first draft of *The Classic of Tea*, the world's first monograph of tea.

In 769, he went to pick Yuejiang Tea in Shaoxing and supervised the processing of tea. By then he was so popular that he had already got young pupils learning from him the skills of tea-savoring.

In 775, taking advantage of the large quantity of information he formerly acquired when compiling *The Complete Collection of Rhymes* for Yan Zhenqing, the prefect of Huzhou, Lu Yu made major modifications on his original manuscript of *The Classic of Tea*, especially the seventh chapter of "Records and Anecdotes", to which he made some amendment and revision. At this point, *The Classic of Tea* became finalized. In the following three years, he wrote *A Contract between Rulers and Subjects* in three volumes, *Origin of Family Surnames* in thirty volumes, *Annals of Figures of the Southern and Northern Dynasties* in ten volumes, and *The Foreboding Year* in ten volumes.

In August of the year 777, when the fragrance of osmanthus blossoms filled the air, Lu Yu, together with some friends, sent Yan Zhenqing off on his trip back to the capital city after Yan left his post in Huzhou. From then on, Lu Yu traveled a lot including Dongyang in Wuzhou, where he called on County Magistrate Dai Shulun, who produced and presented him two poems entitled "Return Poems to Lu Yu the Honorable Reclusive". When winter came he went back to Huzhou.

In 778, Lu Yu moved to Wuxi and met Quan Deyu. Enjoying a sightseeing tour in Huishan, he created "Episodes of Huishan Temple".

In 780, with Jiaoran's support, Lu Yu managed to get *The Classic of Tea* printed. In the same year, he went on a tour to Taihu Lake and visited Li Ye, who generously presented him with his poem of "Delight in Lu's Arrival during My Illness on the Lakeside", showing his gratitude for Lu Yu's visit. And this anecdote gradually became a household story throughout the ages.

By 781, Lu Yu had become more matured in all of his versatile talents and was known to both the court and the public, far and wide. Highly appreciating his remarkable brilliance, Emperor Dezong made an imperial call for him to be the Minister of Education, but he didn't take the post. Soon afterwards, the emperor made another offer assigning him to be the presider in charge of the sacrificial rites in Taichang Temple and again he declined it. So never in his lifetime did Lu Yu hold any public office nor did he get married.

In 783, Lu Yu moved to Shangrao. At a place 2 *li* north of the city, he set out to build a house and pavilions, chisel and channel spring water, and plant tea trees, bamboos and flowers around the residence. Through his intent care and painstaking labor, the new house was completed in the first year (784) of Xingyuan Period. His abode was later called "Hongjian Residence" and the spring, "Shangrao Lu Yu Spring".

In 785, Lu Yu met Meng Jiao in Shangrao. Meng Jiao composed the poem "Inscription for Lu Yu's New Cottage in Shangrao", conveying his congratulations and deep appreciation for the beautiful environment of Lu Yu's residence.

In 786, at the invitation of Xiao Yu, the provincial censor of Hongzhou, Lu Yu came to lodge in the local Taoist temple of Yuzhi. The following year, he compiled a series of poems titled "Poem after Lu Moving to Yuzhi Temple", with Xiao Yu's poem "Delight at Lu's Move to Yuzhi Temple" leading the head and Quan Deyu writing the preface to the collection. It's a pity that the collection of poems has long been lost.

In 789, upon the invitation of Li Fu, the military envoy of Lingnan (Guangdong and Guangxi), Lu Yu went from Hongzhou to Guangzhou, where he met Zhou Yuan, the chief local official. The two of them extended valuable political assistance to Li Fu. Located in the east of the office of the envoy, Lu Yu's house thus assumed the refined name of Eastern Garden and hence Lu Yu was popularly known as Mr. Eastern Garden. The next year, he returned to Hongzhou and continued to live in Yuzhi Temple.

In 792, Lu Yu came back from Hongzhou and resided in Qingtang Retreat, his other abode in Qingtang, Huzhou. Shutting himself away from any disturbance, he was assiduously devoted to his task of writing and in three years, he accomplished *Records of Officials in Wuxing* in three volumes and *Records of Huzhou Prefects* in one volume.

In 794, Lu Yu moved on to Suzhou. He built his house (later called Lu Yu Abode) to the north of Huqiu Hill, where he chiseled a rock well (later called Lu Yu Rock Well and now referred to as Lu Yu Well for brief) for water diversion and tea growing. Meanwhile he wrote his *Evaluation of Springs* in one volume.

In 799, getting on for seventy and feeling a deep yearning for Huzhou, Lu Yu returned to his Qingtang Retreat, outside Qingtang Gate in Huzhou, where he spent his remaining years in happiness and died in around 804.

2. Lu Yu and *The Classic of Tea*

The Classic of Tea, as the world's first monograph of tea, is one of Lu Yu's greatest contributions to mankind. It contains 3 volumes (Volume I, Volume II and Volume III) with a total of ten chapters. The major contents and textual structure are as follows: (1) Origins of Tea; (2) Picking and Baking Tools; (3) Picking and Baking; (4) Boiling Apparatus; (5) Tea Boiling; (6) Tea Savoring; (7) Records and Anecdotes; (8) Producing Regions; (9) Dispensable Tools; (10) Scroll Transcription.

Lu Yu's *The Classic of Tea* contains a set of rules and methods drawn on his observations of tea growth patterns in the various tea-producing regions and of the tea processing on the part of tea farmers, together with his further analysis of the different qualities of tea and his study of the superior ways of tea boiling. In addition, he also devoted special attention to the folk craft of making tea sets and tea utensils so much so that he successfully created and produced a unique set of his own. Lu Yu committed his whole lifetime to the study of tea, leaving his footprints across each and every big tea area all over the country. In 756, at the age of 24, Lu Yu toured with friends every major tea-yielding area, investigating, observing and learning from tea farmers their experience and processing methods. In 760, Lu Yu went back to Huzhou, and, through analyzing and generalizing his collected data of tea study, he undertook the compiling task of *The Classic of Tea*. In 765, Lu Yu completed the first draft of the book based on the data acquired from his field study of the 32 prefectures and on the findings of his years of research. In 780, after decades of utmost efforts and through the ardent assistance of his friends, he accomplished the compilation of the great book and had it ready for publication. The whole book took 27 years to write and the contents, according to Wenyuan Chamber (The Imperial Library) version of *Complete Works of Chinese Classics*, are shown as follows.

2.1 Origins of Tea

Tea trees are a peculiar kind of trees in southern China, which vary in height from one or two feet to even dozens of feet. The regions of eastern Sichuan and western Hubei boast large tea trees with trunks as thick as taking two people to encompass with outstretched arms so that the branches must be cut off for the leaves to be picked. Tea trees are in the shape of *Ilex latifolia thumb* with leaves resembling those of gardenia, flowers those of white rosettes, seeds of palm, stalks of cloves, and roots of walnut.

The Chinese character "茶" for tea may be based on different key radicals for its etymological formation, i. e. , on the part of "grass", on the part of "wood" or on both parts of "grass" and "wood". As for its names, "tea" has the following five different alternatives: *cha*, *jia*, *she*, *ming* and *chuan*.

As to the land quality for tea growth, on top of the list is the soil with fully weathered rock, the second best is gravel soil and loess is the worst.

Normally, mediocre skills produce no fruitful trees and tea seedlings are less likely to be flourishing if they are incorrectly transplanted. The right methods of growing tea are similar to those of planting melons. If they are properly applied, the trees are available for tea picking only in three years.

With regard to the quality, the tea of natural growth in the wild is better than the tea planted in the garden; tea growing on sunny slopes is superior, especially when tea plants nestle in the shade of taller trees, where the purple tea leaves are of higher quality than the green ones. Besides, bud leaves shaped like bamboo shoots are better than the slender ones. In addition, those bud leaves still with curly blade are much better than those already flatting out. Tea trees growing in shady slopes or valleys are of low quality and not worth plucking, for its stagnating property tends to cause abdominal diseases in the drinker.

In terms of the functions of tea, its innate cool and cold nature contributes to the decrease in body internal heat and thus makes it a perfect beverage, especially for those virtuous people with noble temperament and thrifty lifestyle. In dealing with such complaints as fever, thirst, chest distress, headaches, eyestrain, limb weakness and poor joints, tea can be as effective as the legendary nectar and only four or five mouthfuls of it will suffice. However, if the plucking is not timely, or the processing is not fine enough, or the leaves are mingled with weeds or dead leaves, drinking of the tea may get people sick. This is the defect of tea.

In fact, this disadvantage of tea drinking is comparable to the problem with ginseng, whose different origins may result in big quality differences or even adverse effects. The top quality ginseng is produced in Shangdang, medium quality is from Baiji and Xinluo, and lower quality is in Gaoli. Ginseng produced in Zezhou, Yizhou, Youzhou, Tanzhou is of the lowest quality with no curative effect. Or still worse, if the balloonflower root is mistakenly taken as ginseng, no illness whatsoever will be recovered. The awareness of the adverse effects of ginseng is as good as understanding everything about that of tea.

2.2 Picking and Baking Tools

Ying, also called *lan*, *long* or *ju*, is a container like a basket or a cage, woven with thin bamboo strips, with the varied capacity of five liters or one, two or three *dou*. People carry the basket for tea plucking.

Zao is a stove without a chimney.

Fu is kind of a parching wok with a smooth bending at the edge.

Zeng is a wooden or earthen steamer, upright, jointing the wok (under it) with mudding to confine the steam. Inside of it is a bamboo grate (to separate the water from the tea) with shim ears (for easy handle to carry the grate out). To start the steaming of the tea, put the tea on the grate till it is properly steamed, and then take the grate out with the tea on it. If the wok is boiled away, inject into it more water through *zeng*. Use a three-tined branch to stir and unclench the steamed shoots and leaves to avoid the loss of leaf juice.

Chujiu, also called *dui*, is a stone pestle used for mashing steamed tea leaves. The more frequently it is used, the more handy.

Gui, also called *mu* or *quan*, is an iron model or mold used for compressing tea into a certain shape. It may be either circular, square, or in the shape of a flower.

Cheng, also called *tai* or *zhen*, is a stand or an anvil used as a hammering block to cushion against, usually made of stone. However, with the wooden material from pagoda trees or mulberry trunks, it is advisable to have its lower half buried into the ground to make it as unshakable as possible.

Yan, also called *yi*, is made of oil silk cloth or from the worn-out light clothing or a raincoat. Put *yan* on *cheng* and *gui* on *yan* to make compact tea. *Yan* makes it easy to take the compressed tea cakes out of *gui*.

Bili, also called *yingzi* or *panglang*, a bamboo plate or rack with a net mat. It needs two three-*chi*-long bamboo sticks, with which to make a frame with the length of two *chi* five *cun* and the width of two *chi* and the remaining five *cun* to make the handles. As for the net, it is a square mesh sieve woven with bamboo strips. The tool of *bili* highly resembles the vegetable sieves but for the purpose of tea airing.

Qi, also called *zhuidao*, is an awl with a handle made of solid wood, used to puncture holes in the tea cakes.

Pu, also called *bian*, is a rope woven with bamboo used to make it easier to string tea cakes together for easy handle.

Bei is an underground baking range. Dig a pit in the ground of two *chi* deep, two *chi* five *cun* wide and ten *chi* long. Build a wall of two *chi* in height with bricks or stones and then clay it with mud to make it flat and smooth.

Guan is a whittled stick cut from bamboo, two *chi* five *cun* long, used for putting tea cakes into *bei* for baking.

Peng, also named *zhan*, is a wooden rack shed put on *bei* used for baking tea. It is made up of two layers of shelves with one *chi* apart. When the tea is baked to damp-dry, raise it from the bottom of the frame to the lower level and when it is fully dry, raise it to the upper layer.

Chuan is a fine string to thread tea cakes together on. It is made of bamboo strips in Jiangdong (the east of the Yangtze River) and Huainan (the south of the Huaihe River) whereas in Sichuan area, of tough pieces of bark. In Jiangdong a string of tea cakes weighing one *jin* is called a large *chuan*, half a *jin* is a medium *chuan* and four or five *liang* is a small *chuan*. In Sichuan area, a string of 120 *jin* is a large *chuan*, 80 *jin* is a medium *chuan* and 50 *jin* is a small *chuan*. As for the character of *chuan*, it was formerly used as a noun and written the way as in *chaichuan* or a verb as written in *guanchuan*. But now it is no longer the case. The same character of *chuan*, despite its references to both a verb and a noun, assumes respective tones in the two different cases. It is read, like five other characters of *mo*, *shan*, *tan*, *zuan*, *feng*, with a falling tone when referring to a noun while assuming a level tone when denoting the verb. *Chuan*, in this context, is a noun in the falling tone to mean a unit of a string.

Yu is a storage tool (for baked tea cakes) structured with a wooden frame, the peripheral wall is woven with bamboo slices and properly papered with a door opened ajar. Inside, there is a partition to divide the space into two parts. On the top is a lid and underneath is a base tray. In the center is a utensil containing charcoal ashes glowing without flame. During the musty plum rain season in the southeast, fire can be added to it to get rid of the dampness.

2.3 Picking and Baking

Tea leaves are commonly plucked between the lunar months of February, March and April. Bud leaves as strong as bamboo shoots usually grow in fertile soil with weathered rocks. Those new sprouts with length up to four or five *cun*, bearing some resemblances to the tender bines of vetches and ferns breaking through the soil, deserve dewy picking in the morning. In the surroundings of bushy grass bud leaves tend to be shorter and thinner; if there are three to five new tips on a twig, choose and pick the best one that grows tall and straight. It is not advisable to pick on a rainy day or on a bright day with clouds. Pick on a sunny day. The picking should be immediately followed by steaming, mashing, molding, baking, stringing and sealing in order to keep the baked tea cakes dry.

The tea cakes may assume varied shapes and hence present different quality levels. Roughly speaking, some are as creasy as the shrinking leather of the boots of the Northern *Hu* people (in ancient China); some bear the slight wrinkles and folds in the breast of zebu; some resemble clouds afloat above the mountain, coiling and rolling; some look like a breeze creeping over, stirring swaying ripples; some are comparable to the smooth and moist surface of mud paste subsiding from the fine dirt a potter has sieved in water; some are like a newly cultivated land full of ups and downs in the wake of a sweeping torrent. Those kinds are all ranked among tea of the top quality. Conversely, if the tea leaves resemble bamboo shoot shells, with hard stalks not easily steamed or mashed, then the tea cakes are likely to take the shape of a sieve; if the tea leaves assume the appearance of the lotus leaves which are withered by frost, then the tea cakes tend to be shrinking and stale. These sorts are the coarse tea of inferior quality.

There are seven working procedures all the way from the picking to the sealing, and eight quality grades of tea ranging from the boot resemblances to the frosted lotus kind. If you regard the brightness, darkness, flatness and straightness as the measurement criteria of superior tea, your identification is not flawless. In the same way, if you take the yellow, shrinking and uneven features as indicators of good tea, your judgment is as inferior. The true and best identification lies in the abilities to point out both the merits and the demerits of tea. Why? It's because the tea whose juice has already been pressed out tends to be bright; otherwise, shrinkage results. On the other hand, tea processed over the night is dark in color while that made during the day is yellowish. What's more, tea properly steamed and tightly pressed tends to be even and smooth; otherwise, hasty procedures make uneven surface of tea. This kind of tea is nothing better than grass leaves. To tell superior tea from the inferior kind, it is necessary to keep a set of identification tips ready for use.

2.4 Boiling Apparatus

Wind furnace (ash bearing); *ju*(basket woven with bamboo splits); *tanzhua* (charcoal-breaking stick); fire tong; caldron (boiling pot); cauldron stand; clamp; paper bags; grinding groove; sieve and box; *ze*(measuring spoon); water tank; filter pouch; gourd ladle; bamboo chopsticks; salt container; boiled water jar; tea bowls; bowl basket; cleaning brush; washing tank; dross basin; washing towel; utensil cabinet; deposit basket.

The wind furnace (ash bearing), cast in bronze or iron, resembles the ancient *ding*. Its wall is 0.3 *cun* in thickness, the edge (around the upper mouth) is 0.9 *cun*, the hollow inside of the stove is 0.6 *cun* and coated with mud. The furnace has three dumpy legs for support, cast with a total of 21 ancient characters. On one prop foot are 7 characters of "water above, wind below and fire inside", on another foot are 7 characters meaning "balancing five elements to cure all ailments", and on the last one are 7 characters of "cast the year following Tang conquering Hu". Between the intervals of the three legs are three openings of louvers. One more opening underneath the furnace is for ventilation and ash leakage. Above the three windows are 6 ancient characters in all, respectively written as *yigong*, *genglu* and *shicha*, put together to mean the so-called "soup of Yigong and tea of Lu".

On the inside top of the furnace chamber are three buttresses to support the brewing pot. Between the buttresses there are three grids with three patterns respectively. On the first is a figure of a phoenix, a drawing of *li* divination; on the second is a picture of a tiger, which is a symbol of *xun* divination; on the last rib is a picture of a fish, i. e., a water animal, a diagram standing for *kan* divination. None of the three functions is dispensable, *li* for fire, *xun* for wind, and *kan* for water; the wind is an essential element to the fanning of the fire, which in turn is necessary to the boiling of water, hence all the three diagrams. The stove is decorated with designs of plants, water and character patterns. The stove proper can also be made of cast iron or clay. The ash receptacle is a three-footed iron plate used to catch and hold the ash leakage from the furnace.

Ju is a basket woven with bamboo splits, 1.2 *chi* in height and 0.7 *chi* in diameter. Or it may also be rattan woven. To start with, make a wooden frame with the shape of *ju*, and then weave the outside wall layer with bamboo slices or vines in the pattern of six-ribbed round figure, i. e., a hexagonal-shaped circle pattern. When the bottom base and the top cover are folded and fixed on to the cabinet, it takes the shape of a bamboo box. Use small bamboo sticks to whipstitch the sides for decoration.

The charcoal-breaking stick, called *tanzhua*, is used to break charcoal, with usually a six-ribbed iron bar of one *chi* long, one end comparatively thinner and the middle section thick. To handle it, hold on the thin end, to which a small ring is often attached as an ornament. It is somewhat similar to the soldiers' clubs in the Helong area (Gansu Province). The stick can also be made in the shape of either a mallet or a hatchet, or whatever.

Fire tongs are also called *zhu*, kind of chopsticks, in the shape of cylinder of 1.3 *chi* with a flat head and no ornaments or attachments of any kind, usually made of iron or wrought copper.

A caldron, a kind of a boiling pot, is made of pig iron. Now some specialized blacksmiths make pots with the so-called *jitie*, i. e., the iron cast from waste coulter. The inside of the pot is molded with clay and outside of it, with sand. The mud clay keeps the interior surface of the pot smooth and easy to scour its rust off, while the coarse sand outside makes its exterior surface rough and easy for the absorption of flame heat. Pot ears are made in the shape of square because "square" implies righteousness and integrity; pot rim stretches wide for "wide" takes the meaning of expansiveness. The depth of the belly-button section keeps the water boiling in the center, which makes it easier for the tea foam to rise, which in turn purifies the mellowness of tea soup. In Hongzhou pots are made from porcelain and Laizhou, from stone, both of which are elegant-looking vessels but are not so solid or durable in texture. Pots of silver are the cleanest but are too pompous and extravagant. Elegance and cleanness have nothing wrong in themselves but for the permanent use of the pot, iron is irreplaceable and never fails.

The cauldron stand, is a crisscross frame with the middle part hollowed out to prop the cauldron.

The clamp is made of small, green bamboo, 1.2 *chi* in length. Ensure that there is one *cun* between its short end and the joint. Split the long end of bamboo into halves to clip the tea cakes in baking. When the bamboo juice is dripping on the fire, the fragrance of the bamboo is borrowed to enhance the taste of tea. This demand is less likely to be met except in the woods and so, more often than not, such materials as refined iron or wrought copper are more preferable due to their greater durability.

Paper bags are sewn up with double layers of white and thick paper which is made from Shanxi rattan peel, for the storage of baked tea without losing its aroma.

The grinding groove is preferably made of orange wood or less ideally, wood from such trees like pear, mulberry, tung or cudrania. The inside of the groove is round and the outside is square. The circular interior shape makes it easy to roll and the square exterior frame prevents overturning. The inside room of the groove is exactly as spacious as to accommodate the grinding wheel with no extra space. The grinding wheel is shaped like a carriage wheel without spokes and centered by an axis of 9 *cun* long and 1.7 *cun* in diameter. The wooden grinding wheel is 3.8 *cun* in diameter with a thickness of 1 *cun* in the center and 0.5 *cun* at the edges. The axis has a square hole in the middle and the handle is cylinder. The tea whisk for dusting the tea is made of feathers.

The sieve and box is a set of two tools. The tea screened out of a sieve is stored in a box with the cover closed, together with the measuring spoon inside. To make the sieve, bend the splits from big bamboo sticks into round shape, and then attach to the bottom a bolting cloth of yarn or silk. The box is made of a bamboo section with a joint or made by bending fir sheets into the shape of a cylinder and painting the frame. The box is 3 *cun* in height, with 1 *cun* of the lid height plus 2 *cun* of the bottom base height, and 4 *cun* in diameter.

The measuring spoon, the so-called *ze*, can be some kind of easily found replacements like seashells, clams and others. Or it can be made of such materials as copper, iron, bamboo slips and so on. The character of *ze* is literally meant for measurement, criterion, and limit. As a rule, one liter of water requires boiling with one spoon (i. e. ,1 square *cun*) of tea powder to produce a desirable flavor. Still, those favoring weak tea may somewhat reduce the amount while those preferring a stronger taste can add slightly more to it and hence the worthy name of *ze*.

The water tank is made by piecing together boards of such wood like Chinese scholar tree and catalpa. Paint the joints inside and outside with lacquer paste. It has a water capacity of ten liters.

The filter pouch is a utensil for filtering tea dust. As commonly used, its frame is cast with pig copper to prevent producing verdigris, contaminants or foul smell in case of being moistened. If wrought copper is used, it will breed verdigris and contaminants. Or if it is made of iron, the pouch will also smell unpleasantly rusty. Chances are those living in the mountains and woods use bamboo or wood as an alternative, which is not adequate for permanent use or long-distance travelling. Therefore, pig copper is the best choice. As for the making of the pouch itself, first weave a sheet with green bamboo strips and roll it into the shape of a bag. Then, sew a piece of green silk onto the frame and embellish it with a green stone. Next, make a bag out of a piece of green oilcloth and put the whole of the water pouch into it. The water bag has a diameter of 5 *cun* and a handle of 1.5 *cun*.

A gourd ladle is also called a scoop, made from a split gourd or simply dug out of a block of wood. In his *Ode to Tea*, Du Yu of the Jin Dynasty, mentioned "scooping with a gourd ladle". By a "gourd ladle", it refers to a scoop with a wide mouth, thin body skin and a short handle. During Yongjia Period, Yu Hong from Yuyao went to pluck tea in Waterfall Mountain and met with a Taoist monk, who said to Yu Hong, "I'm Danqiuzi. Will you please give away some of your tea to me another day when you have some more in your bail?" The character denoting "bail" is also a wooden scoop, usually dug out of pear wood.

Bamboo chopsticks are tongs made of bamboo or wood from peach, willow, cattail, or persimmon, with a length of one *chi* and two ends coated in silver.

The salt container is usually a round vessel made of porcelain, four *cun* in diameter. Some may also take the shapes of a box, a bottle or an urn. It is used for storing coarse salt flakes. Its *jie*, 4.1 *cun* long and 0.9 *cun* in width, is something made of bamboo and used as a spoon to ladle salt. Literally speaking, *jie* means a polished piece of bamboo.

A jar for boiled water is used for keeping hot water, made of porcelain or sand, with a capacity of 2 liters.

Tea bowls made in Yuezhou (in Zhejiang) are of higher quality than those in Dingzhou and Wuzhou; those from Yuezhou (Hunan) are also superior to those from Shouzhou and Hongzhou. Some people believe teacups produced in Xingzhou are better than those in Yuezhou (in Zhejiang), but it's not the case. If porcelain from Xingzhou is like silver, then that from Yuezhou (in Zhejiang) should be like jade, and this is the first reason for the inferiority of porcelain from Xingzhou. If Xing porcelain resembles snow, then Yue porcelain resembles ice, which is the second reason that Xing porcelain is not as superior. Xing porcelain is pale and makes tea look red, while Yue porcelain is cyan and provides the tea a green color, which is the third reason for Xing porcelain not to be as fine. In the Jin Dynasty, Du Yu in his *Ode to Tea* remarked that "the choiced porcelain vessels are produced in Dong'ou", namely, Yuezhou in Zhejiang. It again shows that porcelain from this place is of top quality. A tea cup from Yue porcelain doesn't have its upper edges curled outward but the bottom loop curled slightly, with a capacity under 0.5 liter. Porcelain from both Yuezhou (in Zhejiang) and Yuezhou (in Hunan) is cyan and contributes to the rosy and milky color of tea. However, the white porcelain from Xingzhou produces a red color of tea, Shouzhou's yellow porcelain lends the tea a purple color, and Hongzhou's brown porcelain makes tea dark. Therefore, porcelain from all these places is inadequate for teacups.

The bowl basket is made by twisting and weaving wisps of cattail. It can hold 10 tea bowls. Or alternatively it can be replaced with a bamboo basket. As to the paper kerchiefs for wrapping and separating the cups, sew two layers of the Shanxi paper into 10 kerchiefs in square shape.

The cleaning brush is made in two ways. First, bundle slices of palm tree hemp with a dogwood branch. Second, cut off a length of bamboo pipe and thrust the bundle of palm hemp into the pipe to form something like a huge brush.

The washing tank is used for holding cleaning water, made by putting together pieces of catalpa wood in the same way that a water tank is made, with a capacity of 8 liters.

The dross basin is used for collecting what is left of the tea and its dregs. The method of its making is the same as the water basin, with a capacity of 5 liters.

The washing towel is made of a piece of 2-foot-long coarse cloth. It usually takes two pieces for alternative use in cleaning the tea sets and vessels.

The utensil cabinet is made of pure wood or of pure bamboo, or of both in the shape of a bed or a case, often painted in black or yellow colors. Some are made with a small door which can be easily locked. Usually the chamber is 3 *chi* long, 2 *chi* wide and 6 *cun* high. As its name *julie* suggests, every kind of tea vessels can be contained and properly displayed inside.

The deposit basket is thus called due to its capacity of holding all kinds of storage appliances. The inside layer is woven with bamboo splits in a triangle-formed square eye shape. The outside layer is knit with both broad and narrow splits, with the broad ones in double pieces and the narrow ones in single pieces, in the pattern of crisscross with the double-splits woven lengthwise and single-sprits filling them crosswise, forming square holes to add some exquisite elegance to the basket. The basket is 1.5 *chi* high, 2.4 *chi* long and 2 *chi* wide. The bottom of the basket is 2 *cun* high and 1 *chi* in width.

2.5 Tea Boiling

Take care not to roast the tea on the fire of ventilation because the flickering flame can be as sharp as a drill and results in uneven heating. When roasting, keep the tea cake as close to the fire as possible and turn it over repeatedly. Heat it until the raised spots on its surface look like the pimples and lumps on the back of frogs and then remove the cake 5 *cun* away from the fire. When the curled tea spreads and becomes flat again, repeat the above roasting procedures. The length of time for roasting depends on the previous ways of drying the cakes in their making. If the cake is formerly dried through roasting, then it should now be roasted until it gives out a fragrant vapor; if the cake is dried in the sun, it is supposed to be roasted until it becomes perfectly soft.

But there is still a third case. At the very beginning of the cake making, when the tender tea leaves are pounded immediately after being steamed, the leaves themselves are easy to be mashed but the stalks remain to be as hard. Some bull-headed efforts or even a powerful pestle of a thousand *jun* (one *jun* equals to 15 kilograms) would still fail to mash them. This is just like paint-coated beads, small and slight as they are, it's impossible for a vigorous man to control and hold them. The properly mashed tea seems to have no stalks and its cake, when roasted, feels as soft as a baby's arm. This is the sign and stage of stopping the baking. Lose no time to store the well-roasted cake in paper bags to prevent the fragrance dispersing and spreading. When the cake cools down, grind it into powder.

As to the fire for the roasting, charcoal is better than strong firewood. However, if the charcoal has been used for roasting meat and has caught the smell of grease, or if the firewood is sooty or obtained from the used and decayed wood products, in neither case is it proper for roasting tea. Ancient people used to believe that wood from broken wooden products will produce a strange taste in the food if it is used for cooking and boiling. They had something there.

Regarding the water for tea boiling, the best water is taken from mountains, the second best is from rivers, and the third is from wells. In mountains, it is most desirable to fetch the slowly dripping water off stalactites, while swiftly surging water is not fit for drinking because its long-term use will cause neck diseases. Water from valleys where several streams converge and stop appears to be clear or even transparent but factually, it is stagnant and may have dragons lurking in and poisoning the water between the times of hot season and First Frost. Therefore, the safe and common practice is, before using the water, to dig a hole and let the filthy, polluted water out so as to keep the new, clean water murmuring in. For the river water, it is necessary to go and fetch from the section far away from residence. However, well water is more drinkable to fetch from the area where more people reside. When water is boiling, there are bubbles popping up like fish eyes with slight sound, which is known as a first boiling. When bubbles keep coming up like beads at the edge of the pot, it is a second boiling. If the water continues boiling with lumpy waves, it is a third boiling, the point when prolonged boiling would spoil the taste of the tea. As soon as the first boiling starts, add a proper amount of salt to flavor the water and give away what is left of the tasted water. Take care not to continue adding salt merely due to its treacherous tastelessness; otherwise, it would seem as if saltiness were the only adorable taste. At the second boiling, scoop out a ladle of water. Stir the remaining water in a circle with bamboo chopsticks, measure a proper amount of tea with a measuring spoon and pour the tea in around the center of the whirlpool. Soon after, when the water is boiling with foams splashing, add the water formerly ladled out to stop the boiling and keep the "essence" generated on the water surface.

To serve the tea, scoop tea into bowls, keeping the froth and foams evenly distributed, which are the essence of tea. The thin kind of it is called the spray and the thick kind is called foam. The light kind is called flowers, like date blossoms floating on the circular pond, or a pool of water winding and zigzagging forward, or duckweed newly emerging among oasis, or scales of clouds floating in the bright sky. Its froth looks like moss afloat on the edge of water or chrysanthemums fluttering into the cup. As for the foam, it is the thick snow-white layer of bubbles piled up upon the boiling of water when the dross is being cooked. No wonder *Ode to Tea* describes it as "bright as snow and splendid as spring flowers" and it is rightly the case.

When water is boiling for the first time, skim the film of black mica off the water surface, which would spoil the taste of tea. And after that, the first scoop ladled out is permanently fragrant, which is usually kept in the boiled water jar for later use to stop boiling and keep the "essence". Hereafter, the following first, second and third cups of tea taste slightly less fragrant but are far more palatable compared with the subsequent fourth and fifth. Beyond that, the rest is not suitable for drinking except to quench extreme thirst. Generally, one liter of water can be evenly divided into five cups and the tea is supposed to be drunk in successive sips while it is hot, during which the dregs are condensed at the bottom and the essence is afloat above. Once it cools down, the essence will disperse with the vapor. The same is true to the remaining tea soup in the boiling pot.

Tea has the nature of "thrift", which requires a limited quantity of water to boil with; otherwise, an excessive amount would diminish the taste of tea. As is often the case, the taste does not feel as strong even drinking halfway into a bowl of tea, let alone when more water is added. The right color of tea soup is pale yellow, with exquisite fragrance spreading all around. Tea with a sweet taste is *jia*, tea with a bitter taste overwhelming sweetness is *chuan*, and tea that tastes bitter when in the mouth but sweet when being swallowed is *cha*, which is tea in its true sense.

2.6 Tea Savoring

As the laws of nature have it, birds are endowed with wings and the innate ability of flying, beasts with fur and the capacity of running, and man with mouth and the faculty of speaking. Despite the differences, born between the heaven and earth, they all rely on water and food for their existence. Therefore, drinking is significantly important. To quench thirst, people need water; to dissipate grief, people need alcohol; to lift drowsiness, people need tea.

As a drink, tea started from Shennong. It is, however, the records by Duke of Zhou that make it known to all. Many people enjoyed tea drinking, such as Yan Ying from the Qi State in the Spring and Autumn Period, Yang Xiong and Sima Xiangru in the Han Dynasty, Wei Yao from the Kingdom of Wu in the Three Kingdoms Period, Liu Kun, Zhang Zai, Lu Na, Xie An, Zuo Si and others in the Jin Dynasty. Later, tea was increasingly popular and even gradually becoming a vogue before it reached its peak in the Tang Dynasty, when tea was to the liking of every household in the capital cities of Xi'an and Luoyang, as well as in the areas of Jing and Yu and so forth.

There are such types of tea as coarse tea, loose tea, powder tea and caked tea. To make a tea beverage, the newly picked tea leaves must go through the few procedures of chopping, steaming, baking and grinding. Some people easily make the so-called "raw tea" simply by putting the tea in a jar and pouring in freshly boiled water. Others choose to boil the tea with green onions, ginger, dates, tangerine peels, cornels, mints and the like; after boiling it for a prolonged time, they either lift the pot to make the tea clear or skim off the foams. Unfortunately, tea thus made is nothing better than the waste water poured into the ditches. However, it is customary to have it done this way.

Ah, all natural things are created with their own supreme subtleties. However, what people are adept at is only what is of supreme simplicity. They inhabit splendidly decorated houses, wear magnificently tailored attires, eat delicately prepared food and drink tastefully brewed liquors. They are, however, not good at tea making. After all, there are generally nine demanding aspects concerning tea matters: first, the plucking; second, the identification; third, the utensils; fourth, degrees of heating; fifth, water quality; sixth, the baking; seventh, the mashing; eighth, the boiling; and ninth, tea-tasting. It is not proper to pluck tea on a cloudy day or to bake tea during the night; it's not the real identification simply trying to know the flavor by chewing or learning the fragrance by sniffing; it's not the right vessel to use if it smells of mutton or fish; it's not the desirable fuel if taken from lampblack wood or the charcoal that has been used for roasting meat; it's not the right water to use if it is from the swift surges or a stagnant pool; it's not the right way of baking if the outside is over-done while the inside is not yet well-done; it's not appropriate if the tea is mashed into kind of fine green powder or yellowish dust; it's the lack of skill that leads to hasty stirring; it's not the proper manner to drink tea only in summer and stop doing it in winter.

Tea soup, which can be ranked as the most deliciously fragrant beverage, is so precious that each boiling can only provide three bowls; otherwise, at the sacrifice of the high grade flavor, each boiling can offer five bowls at most. If there are five people drinking tea, scoop out three bowls to pass around among them; if there are seven people, ladle out five bowls to go around; if there are six people, there is no need to bother about the missing amount of the sixth bowl because what is lacking in amount can be compensated for by the permanent flavor and the lingering essence of the five cups of tea.

2.7 Records and Anecdotes

Historical figures related to the tea are shown as follows. Emperor Yan, namely, Shennong, one of the ancient Three Emperors; in the Zhou Dynasty, Duke of Zhou from the Lu State and Yan Ying, the prime minister of the Qi State; in the Han Dynasty, immortals of Danqiuzi and Huangshanjun, Sima Xiangru the royal gardener, and Yang Xiong the warrior with a halberd; in the Three Kingdoms Period, Sun Hao the dismuke and Wei Yao the top preceptor from the Wu Kingdom; in the Jin Dynasty, Sima Zhong (Emperor Hui), Liu Kun the minister of public works and his brother's son Liu Yan the prefect of Yanzhou,

Zhang Zai the closest emperial bodyguard, Fu Xian the military official, Jiang Tong the high counselor of Prince Huai, Sun Chu the military inspector, Zuo Si the chamber overseer, Lu Na from Wuxing and his brother's son Lu Chu the satrap of Kuaiji, Xie An the general, Guo Pu the satrap of Hongnong, Huan Wen the prefect of Yangzhou, Du Yu the secretary, Fayao the monk from Xiaoshan Temple in Wukang, Xiahou Kai of Peiguo, Yu Hong from Yuyao, Fu Xun from Beidi, Hong Junju from Danyang, Ren Yuchang from Lean, Qin Jing from Xuancheng, Shan Daokai from Dunhuang, Chen Wu's wife from Shanxian County, an old woman from Guangling, Shan Qianzhi from Henei; in the Northern Wei Dynasty, Wang Su from Langya; in the Song Dynasty of the Southern Dynasty, Liu Ziluan the prince of Xin'an and his brother Liu Zishang the prince of Yuzhang, Bao Zhao's sister Bao Linghui, Tanji the monk in Bagong Mountain; in the Qi Dynasty of the Southern Dynasty, Emperor Wu; in the Liang Dynasty of the Southern Dynasty, Liu Ran the main judicial officer and Tao Hongjing the noted scholar; in the Tang Dynasty, Xu Ji the official contributing remarkably to the establishment of the Tang Dynasty.

The historical records about tea are listed as follows.

According to *Treatise on Food* by Shennong, "Long-term drinking of tea makes people energetic and sagacious."

Duke of Zhou remarks in *Erya* that "*Jia* is a kind of bitter tea."

Guangya has it that "In the areas of Jingzhou and Bazhou, people often make the newly plucked leaves into tea cakes. If the cakes are made of old leaves, it is advisable to dip them into rice paste to be coated a little bit. To make the tea drink, first roast the cake till it turns to a reddish color, mash it into powder and put the powder in a porcelain utensil, then pour in hot water. Some others boil the tea together with green onion, ginger, or orange to flavor its taste. This kind of tea beverage has the function of dispelling the effects of alcohol and keeping people away from drowsiness."

Spring and Autumn Annals of Yan Ying keeps the record that "When Yan Ying was serving as the Prime Minister of Emperor Jinggong of the Qi State, what he enjoyed was nothing more than millet rice and roast birdmeat and a few eggs plus tea and vegetables."

Sima Xiangru in his book of *Fanjiang Pian* ranked tea (*chuancha*) among herbs, enumerating them as the following: "lateral roots of *Aconitum carmichaeli*, root of *Platycodon grandiflorum*, bud of *Daphne genkwa*, tender petiole and bud of *Petastesjaponicus F. Schmidt*, *F. cirrhosa* or *Fritillaria thumbergii*, *Phellodendron amurense Rupr.*, *Piper betle L.*, *Scutellari a baicalensis Georgi*, tuber of *Paeonia lactiflora*, *Osmanthus fragrans*, dried root of *Rhaponticum uniflorum*, *Carduus crispus L.*, *guajun* (similar to fungus), *chuanch* tea, tuber of *Ampelopsis japonica*, root of *Angelia anomala* or *dahurica*, dried root of *Acorus calamus Linn*, mirabilite, *guan* peper, dogwood."

In *Dialects*, Yang Xiong from the Han Dynasty noted, "People in the southwestern Sichuan call tea *chuan*."

From *The Chorography of the Wu Kingdom · Biography of Wei Yao*, an excerpt goes, "Every time Sun Hao the dismuke held a big banquet to treat his followers, he would set the minimum limit of 7 liters of alcohol whatever their liquor capacity. However, he showed special partiality to Wei Yao, who had only a capacity of no more than 2 liters, and secretly offered him tea to substitute for alcohol."

Another anecdote narrated in *The Book of Zhongxing in the Jin Dynasty* is about Lu Na: "When Lu Na served as the satrap of Wuxing, Xie An, a general of the guardian force, was once coming for a visit. Lu Na's nephew Lu Chu noticed that Lu Na hadn't made any preparations and, afraid to ask about it, he prepared in private a dinner huge enough for more than 10 people. Upon Xie's arrival, seeing that what Lu Na offered to entertain his guest was nothing but some tea and fruits, Lu Chu set out his abundant, delicious food to treat him. After the guest's departure, Lu Na flogged Chu 40 strokes with a stick and said, 'It's all right and well beyond reproach if you cannot give credit to me. But how come you should defile my integrity and uprightness supposed to be maintained for my whole lifetime?'"

The story goes in *History of the Jin Dynasty* that "Huan Wen, the prefect of Yangzhou, was frugal by nature. Every time he entertained guests, he would prepare nothing more than seven plates of fruits to go with some tea."

According to *Anecdotes about Spirits and Immortals*, "After Xiahou Kai died of disease, one of his clansmen by name of Gounu saw his ghost coming back for his horse, resulting in his wife's sudden sickness. Kai was wearing a flat cap and thin clothes, sitting on the big bed near the western wall that he used to sleep on when alive, and asking for a drink of tea."

Liu Kun in his *Letter to My Nephew Yan, the Prefect of Southern Yanzhou* says, "The other day I received your parcel containing a *jin* of dried ginger, a *jin* of asccharum, a *jin* of radix scutellariae, all of which are exactly what I need. I am not in good shape, frequently suffer from the fidgets and have to depend on tea to refresh my mind, so I long for some authentically super tea for my fitness. Please get and send me some."

Fu Xian's *Instructions from an Inquisitor* quotes the following story: "It is said that in the south an old lady from Sichuan selling tea porridge got herself into great trouble because the local officials, in order to uphold the imperial edicts of industry and thrift, banned her from doing so and furiously broke her tea vessels. Later on they allowed her only to sell tea cakes in the market. How can they be so harsh as to baffle a granny who is trying to sell tea porridge?"

In *The Book of Weird Miracles*, a story goes like this, "Yu Hong from Yuyao went into the mountains to pick tea leaves and came across a Taoist monk leading three black oxen, who took Yu Hong to the Waterfall Mountain and said, 'I am Danqiuzi. I hear you are fond of collecting tea utensils and drinking tea; I have for long wished you could present me with some tea. There are many large tea trees which I can send you as a gift. Someday when you have some more tea in your bowl, please offer some to me.' So when he got home, Yu Hong built a memorial hall to worship the immortal master. Later on, he frequently arranged for his families to search the mountain and, as expected, found the claimed enormous tea trees."

Zuo Si in the Jin Dynasty had a poem titled "My Cute Daughters" with the following description: (Adopted translation from Jiang Xin: P57)

My two daughters are cute girls,
Fair and flawless as lily pearls.
We give the younger the name Pure,
Her tongue's glib but never demure.
The elder's name is an orchid fine,
Brows are rainbow and eyes shine.

They brisk in woods like two fairies,
Can't wait to get ripe fruits and berries.
To flowery nature they're so much bound,
Wind and rain chorus a cheerful sound.
Tea scents from home lure them with desire,
Pursing rosy lips they help blow the fire.

In his poem of "Stepping on Baitu Building in Chengdu", Zhang Zai exclaims:

Where is the bygone Yang Xiong's abode?
Nowhere is the charm of Sima Xiangru's mansion to be showed.
The extravagant richness of Cheng and Zhuo
Matched the abundant luxuries of the imperial household.

Outside on horseback are guests streaming,
Jade ribbons and gold swords across are hanging.
Inside are delicacies from land and sea,
With ready flavors uniquely enticing.

Beyond the mansions in the distant view,
Boundless woods shine with autumn fruit,
Fish abound in the spring river tender and fat,
With carps and oranges the heaven's best food.

Outshining all other kinds of drinking is Sichuan tea,
Spreading afar is its aroma and prestige,
The city gratifies anyone to come here
With a life of rich happiness and peace.

Fu Xun's book of *Seven Instructions* describes some local specialties including tea: "peaches from Shanxi, crab apples from Henan, persimmons from Shandong, chestnuts from Hebei, yellow pears from Hengyang, tangerines from Wushan, tea from Nanzhong and the jaggery from Western Regions."

As Hong Junju says in his *On Food*, "When meeting guests, after the greeting, it is customary to serve them fragrant tea with white foams afloat. Three cups later, there will be an offer of a small cup of juice of the following kinds: sugar cane, *Chaenomeles sinensis*, plum, waxberry, *Schisandra chinensis*, *Canavium album Raeuseh*, raspberry, and *Malva verticillata*."

In his poem Sun Chu enumerates some local specialties including tea:

Dogwood fruit grows on the beautiful tips of the trees,
In the water of the Luoyang River savory carps breed,
Hedong is where snow-white salt is produced,
Shandong is the cradle of delicious fermented soybeans.
Ginger, cinnamon, and tea are special products of Sichuan,
Pepper, orange, and magnolia come where high mountains stand,
Smartweed and perilla abound in the edge of the trench,
People reap fine polished rice in the fertile land.

Hua Tuo remarks in his *On Food* that "Drinking bitter tea regularly is good for mental activity."

Dietetic Restraints, written by a Taoist Hermit Pot, explains that "Long-time drinking of bitter tea improves physical fitness and makes the drinker slimmer and lighter. However, drunk with leek, it tends to add to the body weight."

Guo Pu says in his *Notes to Erya Dictionary*, "Tea plants stand as short as gardenia trees, with their leaves remaining green in winter. Tea leaves can be plucked and boiled to make a drink. Now tea has acquired different names: the leaves first picked are called *cha* and those plucked at a later time are called *ming* or *chuan*, both of which refer to bitter tea as people in Sichuan call it."

In *A New Account of Tales of the World*, "Ren Zhan, with a style name of Yuchang, enjoyed a good reputation from his youthhood. However, ever since he took refuge in the South, he had lost his memory and became muddle-headed. On one occasion when he was attending a tea party, he asked a fellow guest, 'Is this *cha* or *ming*?' Realizing the strange expression on the other part, he defended himself by saying 'I meant to ask whether it is served hot or cold.'"

In *Sequel to Anecdotes about Spirits and Immortals · Emperor Wu of the Jin Dynasty*, it is recorded that "A Xuancheng native named Qin Jing often went to the mountains in Wuchang to pick tea leaves. Once he met with a hairy man of more than 1 *zhang* in height, who led Qin Jing down the mountain, showed him a wide stretch of bushy tea trees and then left. Shortly afterwards, he returned and took an orange formerly tucked in his bosom and gave it away to Qin Jing. Qin Jing was so frightened that he hurried away, but still had the presence of mind to take his tea leaves home."

Rebellion of the Fourth Prince in the Jin Dynasty gives the following account: "When the fourth prince revolted, Emperor Hui was forced to escape from the capital city. Later upon his return to Luoyang, his imperial secretary served him tea with a porcelain bowl to relieve his thirst and fatigue."

Weird Garden records, "The wife of Chen Wu, from the Shanxian County, was widowed rather young and lived together with her two sons in their old residence. She was fond of drinking tea. There was a tomb in the courtyard. Every time she drank tea, she would offer the first cup to the tomb. Her two sons felt annoyed and said, 'The ancient tomb knows nothing about it and it's a waste of your kindness and goodwill.' They even threatened to dig it out. Painstakingly, the mother managed to discourage them from doing so. Over the night, she dreamed of a man saying to her, 'I've been resting here in this grave for over three hundred years, but the sons of yours are always intending to destroy it. I am lucky enough to have received your undeserved protection and generous offerings of tea, so how do I dare not to repay your kindness despite the mere skeleton as I am in this underworld!' Early the next morning, she got 100,000 copper coins which seemed to have been buried for a long time but the string threading the coins still looked quite new. When the mother told them all about it, the two sons felt guilty and began to join their mother in the sacrifice. From then on, the family worshiped the ancient grave more and more frequently."

Story of a Granny in Guangling goes that "During the reign of Emperor Yuan in the Jin Dynasty, there was an old lady who carried a pot of tea to sell in the market early in the morning every day. People in the market were rushing for the purchase but the tea in the pot never seemed to reduce from morning till night. What money she gained was all scattered to the orphans, the poor and the beggars on the road. Some people felt it strange until officers in charge of the law government (called *facao*) captured her and put her under arrest. Strange enough, when night came, the old lady flew out of the prison window together with her tea-selling utensils."

The book *On Life Arts* records that "Shan Daokai, a man from Dunhuang, was never afraid of the extreme summer heat or winter cold. He often ate alchemic stone. He also took a kind of elixir that gave off the aroma of prairie dock, cassia and honey. Besides, he drank nothing but the beverage of tea and perilla".

Sequel to Biographies of Eminent Monks by a Buddhist Daogai has the following record: "There was a monk named Fayao in the Song Dynasty of the Southern Dynasty, whose secular surname was Yang, and whose natal place was Hedong. During Yuanjia Period, he came down to the south of the Yangtze River and met with Shen Taizhen, who lived in Xiaoshan Temple in Wukang, to which Fayao was invited as a presider. At that time, Fayao was already advanced in age. As a matter of fact, he never had a meal without a cup of tea. During Yongming Period, edicts came from the emperor instructing officials from Wuxing to ceremoniously escort Fayao to the capital city. That was the year when Fayao was already 79 years old."

In *Biographies of the Jiang's Family*, "Jiang Tong, named Yingyuan, a high counselor of Prince Minhuai, often submits written memorials to the throne asserting the fact that 'Now in the imperial Western Garden people are selling things, like vinegar, noodles, baskets, vegetables, tea and the like, which is detrimental to our national dignity.'"

Records of the Song Dynasty of the Southern Dynasty offers this story: "Liu Ziluan, prince of Xin'an and his brother Liu Zishang, prince of Yuzhang, paid a visit to Monk Tanji in Baggong Mountain. The monk served tea respectfully. After tasting it, Zishang said, 'This is, as it were, kind of nectar; why call it bitter tea?'"

Wang Wei's "Miscellaneous Poems" includes the following lines:

Silently and quietly I closed the cabinet window,
In solidarity stands the mansion empty and hollow,
Vain eagerness for your return dampens my heart,
Wish tea to drown my sorrow and worries about you.

Bao Linghui, younger sister of the famous poet Bao Zhao from the Song Dynasty of the Southern Dynasty, wrote " Ode to the Fragrant Tea " to eulogize tea.

Emperor Wu in the Qi Dynasty of the Southern Dynasty wrote in his *Posthumous Edict* the following instructions: " Don't make offerings with slaughtered animals on my altar; it is alright and sufficient just to set some cakes, tea, rice, liquor and dried meat. "

In the Liang Dynasty of the Southern Dynasty, in his "Thanks to Duke of Jin'an for the Bestowed Provisions", Liu Xiaochuo wrote "Li Mengsun reads Your Majesty's edict granting me eight kinds of delicacies like rice, liquor, melons, bamboo shoots, sour pickled cabbages, dried meat, preserved fish and tea. The smell of rice is comparable to that of the best brew of Xincheng; and the liquor is aromatic in smell and mellow in taste. The newly born bamboo shoots at the edges of water are even better than the rare sweet flag and banana-plant; the melons from the fields exceed any kinds of delicacies; deer meat tied with cogon grass tastes good, but it is incomparable to the dried meat you bestowed on me. The canned river carp is no match to your gift of preserved fish. The grains of rice are as crystal as pearls, and the tea looks as shiny as the rice. The mere sight of the sour pickled cabbage is simply appetizing. So abundant is your offer of food that even if I had to go a thousand *li* away I wouldn't need to prepare food any more. I keep undying gratitude to Your Majesty, which is unforgettable in my whole lifetime. "

Tao Hongjing's *Miscellaneous Notes* keeps this narrative: "Bitter tea can keep people light and vigorous. In the past, immortals Danqiuzi and Huangshanjun had the habit of drinking bitter tea. "

The Record of the Northern Wei Dynasty says, " Wang Su, a native of Langya, once served as the secretary prime minister for the Qi Dynasty of the Southern Dynasty. He was fond of drinking tea and of *Brasenia schreberi* soup.

Later he returned to the north and began to take great interest in mutton and milk. Someone asked him, 'What if you compare milk with tea?' He answered, 'By no means can tea be equated with milk.'"

The book of *Herbalist Tongjun's Notes* provides the following description: "In places like Xiyang (now Huanggang in Hubei), Wuchang, Lujiang (now Shucheng in Anhui) and Jinling (now Changzhou in Jiangsu) people like drinking tea. They all have tea ready at hand for entertaining guests. Tea is healthy owing to its froth. Generally speaking, beverage plants, i. e., plants that can make a drink, are usable for such a purpose mostly on the part of their leaves. However, as for the two plants of *Asparagus cochinchinensis* and *Smilax china*, people dig their roots out to make healthy drinks. There is a unique kind of fragrant tea in Badong area, drinking of which after boiling can keep people wide awake the whole night. Folklore has it that people like boiling tea with leaves of *Santalum album linn* and *Rhammus davurica Pall* due to their cool nature and the function of reducing internal heat. In addition, in the south there is a kind of melon *Ilex latifolia thumb*, which resembles tea trees and also has a bitter taste. People pick and make the leaves into powder before boiling and drinking it like tea. This drink can make people sober all the way through the night and so the witted salt traders extremely favor this kind of tea and make it their special beverage to refresh their minds and heighten their spirits. Besides, people in Jiaozhou and Guangzhou are also fond of this kind of tea so that they would never fail to serve it first thing to their guests by mixing it with some fragrant sorts of ingredients."

According to *The Geoscience Notes of Kunyuan*, "In Wushe Mountain 350 *li* northwest of Xupu County in Chenzhou (Hunan) lived a minority people who, on every festival or other big occasions, would gather with other members of the same clan, singing and dancing on the mountain which was covered with exuberant tea bushes."

A Comprehensive Atlas states that "There is a creek for boiling tea called Tea Steam 140 *li* east of Linsui County."

The Records of Wuxing by Shan Qianzhi says that "Wenshan Mountain which is 20 *li* west of Wucheng County (in Huzhou) produces tribute tea."

As shown in *The Topography of Yiling with Illustrations*, "Tea is produced in the mountains (of Yiling County) like Huangniu, Jingmen, Nüguan, Wangzhou and others."

The Topography of Yongjia with Illustrations states that "There is a mountain producing white tea 300 *li* east of Yongjia County."

The Topography of Huaiyin with Illustrations depicts that "There is a slope of tea trees 20 *li* south of Shanyang County."

"Chaling is thus named due to its hills and valleys that are covered with large patches of tea trees," says *The Topography of Chaling with Illustrations*.

Materia Medica · Wood Section has the illustration that "*Ming* is also known as bitter tea, a mixture of bitterness and sweetness in flavor, slightly cold in nature, non-toxic, curative for fistula sores, diuretic, expectorant and antitussive, antipyretic and reducing sleep. Leaves plucked in autumn taste bitter and have the effects of expelling flatulence and promoting digestion. Notes to *Materia Medica* adds the suggestion of 'plucking the leaves in spring'."

Materia Medica · Vegetable Section: "Bitter tea, also known as *tu*, or *xuan* or *youdong*, growing in the valleys, hills, or on the roadside of Yizhou, capable of surviving the coldest winter days, plucked on March 3rd of the Chinese lunar calendar every year and dried in the sun. According to Notes to *Materia Medica*, 'It is supposed to be what we have as tea, also called *tu*, drinking of which makes people sleepless.'" Notes to *Materia Medica* records, "According to *The Book of Songs*, 'Who can call *tu* bitter?' and 'Violet tea is sweet', both of which refer to the bitterness of tea. What Tao Hongjing means by bitter tea is woody plants and not vegetables. *Ming*, if picked in spring, is called bitter tea."

In *Canon of Moxibustion and Acupuncture Preserved in Pillow*, a prescription goes as follows: "To cure unhealed fistula boils of long-time standing, bake tea together with centipede until fragrant, divide them into two, mash and sieve them, apply one portion to the afflicted part after washing the skin with the soup boiled from liquorice."

Prescriptions for Children has such a recipe: "For the treatment of pediatric unprovoked convulsion, boil bitter tea with onion fibrous root to make a soup and drink it."

2.8 Producing Regions

The best tea in Shannan Dao is produced in Xiazhou, next best is in

Xiangzhou and Jingzhou, while ranked last on the list is the tea from Hengzhou, Jinzhou and Liangzhou.

Tea of the top quality in Huainan Dao is best yielded in Guangzhou, and second comes from Yiyang Prefecture and Shuzhou, while tea from Shouzhou is inferior to that of Qizhou and Huangzhou.

Tea in Huzhou is the best in the area of Zhexi, with Changzhou ranking the second and Xuanzhou, Hangzhou, Muzhou, Xizhou ranking the third, while that of Runzhou and Suzhou ranks the last.

As to Jiannan Dao, Pengzhou produces the best tea and Mianzhou and Shuzhou, next to the best, followed by that of Qiongzhou, Yazhou and Luzhou. The tea of Meizhou and Hanzhou is somewhat inferior to that of the previous areas.

In the area of Zhedong, Yuezhou produces the best tea, Mingzhou and Wuzhou, the second best, and Taizhou, the third.

Qianzhong Dao has the best tea in Sizhou, Bozhou, Feizhou and Yizhou.

Jiangnan Dao has the best tea in Ezhou, Yuanzhou and Jizhou.

In Lingnan Dao, Fuzhou, Jianzhou, Shaozhou, Xiangzhou produce the best tea.

Despite the lack of detailed information about the tea produced in the 11 places of Sizhou, Bozhou, Feizhou, Yizhou, Ezhou, Yuanzhou, Jizhou, Fuzhou, Jianzhou, Quanzhou, Shaozhou and Xiangzhou, they all rightfully boast superb tea based on what I, for one, have obtained and tasted.

2.9 Dispensable Tools

As to the tea-processing tools, seven kinds can be dispensable and their corresponding procedures of tea making can be naturally omitted if the picking takes place during the Cold Food Festival in the temples of the wilderness or in gardens of the mountains with many hands participating and if the tea leaves can be immediately steamed, mashed, and baked. The tools are *qi* (the awl), *pu* (bamboo rope), *bei* (baking range), *guan* (whittled bamboo strip), *peng* (two-layer rack), *chuan* (fine string for measurement of tea cakes) and *yu* (wooden box for storage baked cakes).

With regard to the boiling tools, if it is between pines and there are stones to sit on, then the *julie* (utensil cabinet) can be omitted. If dry wood and *ding* (boiling pot) and the like can be used for boiling water, then such things as *fenglu* (wind furnace), *huicheng* (ash holder), *tanzhua* (charcoal breaking bar), *huojia* (fire tongs) and *jiaochuang* (cauldron stand) are all unnecessary. If it is in the upper stream or the spring, then there is no need for *shuifang* (water tank), *difang* (a washing tank) or *lishuinang* (water pouch).

If there are no more than five people and the tea can be ground into fine powder, then *luohe* (screen and box) can be without. If you choose to enjoy the tea on a steep rock or in the cave, you can first bake and mash the tea properly at the mouth of the cave and wrap it in paper or put it into a box, so you can leave out *nian* (grinding groove) and *fumo* (duster). Suppose all the items of *piao* (gourd ladle), *wan* (tea bowls), *jia* (bamboo chopstick), *zha* (washing brush), *shuyu* (boiled water jar) and the salt cruet are all contained in *ju* (bamboo cabinet), then *doulan* (deposit basket) becomes useless. However, in an imperial household in the city, the absence of even one single utensil from the 24 kinds will deprive their good mood for a drink.

2.10 Scroll Transcription

Use white silk material of quadruple or sextuple width to write out the above content respectively, and hang it on both sides of the seat. In this way, the origins of tea, picking and baking tools and methods, boiling apparatus, tea boiling, tea savoring, records and anecdotes, producing regions, and dispensable tools are readily available for you to read and to keep in memory. As a result, the contents of the whole book of *The Classic of Tea*, from start to finish, are ready for your easy access.

3. An Inventory of Tea Books in Huzhou Throughout the Ages

Ever since the foundation of tea science in our country by Lu Yu's writing of *The Classic of Tea* in the Tang Dynasty, the traditional study of tea has achieved gradual development and increasingly reached its maturity. Wan Guoding records in his *Tea Book Inventory and Gist* a total of 98 kinds of tea books that were created during the 1,150 years from the advent of Lu Yu's *The Classic of Tea* through the Dynasties of Song, Yuan, Ming and Qing, including 7 kinds from the Tang Dynasty, 25 from the Song and the Yuan Dynasties, 55 from the Ming Dynasty, and 11 from the Qing Dynasty. After years of verifying plenty of ancient literature and integrating all kinds of resources, we find that celebrities of Huzhou in each dynasty of Tang, Song, Yuan, Ming and Qing were, starting from Lu Yu, compiling books of tea and that there are 16 kinds of tea books, accounting for 16.3% of the ancient tea books nationwide, with 6 kinds from the Tang Dynasty, 2 from the Song Dynasty, 1 from the Yuan Dynasty, 6 from the Ming Dynasty and 1 from the Qing Dynasty. Huzhou is uniquely rare across the country not only with its great number of the ancient tea books, or in terms of the penetration of its tea theory expounding, but also in the richness of its tea culture connotations.

The following is a list of the ancient tea books in Huzhou including their titles, brief introduction to the authors and the present situation of their existence.

Table 1 Ancient Tea Books in Huzhou

Title	Dynasty	Author	Brief Introduction to the Author	For Reference
The Classic of Tea	Tang	Lu Yu	Lu Yu (733- c. 804) is the author of the first tea monograph both at home and abroad, the founder of tea science in our country. *The Classic of Tea* was written in Huzhou and accomplished in about 760.	According to Ouyang Yingming's *Abridged Table to the Versions of The Classic of Tea*, among the 70 kinds of domestic collections of *The Classic of Tea*, 6 kinds have been lost. The 33 kinds of overseas collections are either in America's Library of Congress and Japanese Congress Library or have been collected and included into the Encyclopedia Britannica and the Republic of Korean folk tea culture readings.

To be continued

Title	Dynasty	Author	Brief Introduction to the Author	For Reference
Episodes of Mount Guzhu	Tang	Lu Yu		In the Tang Dynasty, in his Preface to *Miscellaneous Odes to Tea* (it is seen as a prelude to *The Classic of Tea* by the later generations), Pi Rixiu related that "The two writings of *Episodes of Mount Guzhu* are mostly about tea anecdotes." Its existent version is the one excerpted and revised by Zhang Hongyong, according to which *Episodes of Mount Guzhu* includes three tea anecdotes from *The Classic of Tea*. Therefore, we come to the conclusion that the book of *Episodes of Mount Guzhu* was prior to *The Classic of Tea* and was later collected into the seventh part of *The Classic of Tea* and entitled "Records and Anecdotes".
Rhymed Formulas on Tea	Tang	Jiaoran	Jiaoran (730-799), secular surnamed Xie, style named Qingzhou, the presider and poet monk of Miaoxi Temple on Zhushan Mountain in Huzhou. Lu Yu once lodged this temple and made a "monk and layman cross-age friendship" with Jiaoran.	*Rhymed Formulas on Tea* has been lost. According to *Biography of Mr. Fuli* by Lu Guimeng, collected in Volume 796 of *Finest Blossoms in the Garden of Literature*, "I wrote *Book of Evaluating and Ranking*, following *The Classic of Tea* and *Rhymed Formulas on Tea*. According to some notes, *The Classic of Tea* was written by Lu Jici, i. e. Lu Yu; *Rhymed Formulas on Tea* was created by Monk Jiaoran."

To be continued

Title	Dynasty	Author	Brief Introduction to the Author	For Reference
Reviews on Tea	Tang	Pei Wen	Pei Wen, Huzhou Prefect between the 6th year and the 8th year of Yuanhe in the Tang Dynasty, left his stone inscription when supervising tribute tea on Guzhu Mountain.	The book first appeared in the Song Dynasty in Liu Yan's quotation in *The Countermeasures of the Final Imperial Examination*, the 28th volume of his *Collected Works of Longyun*, which says, "Such people as Wen Tingjun, Zhang Youxin and Pei Wen either wrote *Comprehensive Work on Tea*, or *The Classic of Water* or expounded on details of Guzhu Mountain." It seems, from the above records, that Pei Wen's *Reviews on Tea* was a kind of exposition about Guzhu tea. In *The Sequel to The Classic of Tea* in the Qing Dynasty, Writer Lu Tingcan used some excerpts from Pei Wen's *Reviews on Tea*.
A Miscellany of Tea Processing	Tang	Zhang Wengui	Zhang Wengui, Huzhou Prefect from the 1st year to the 2nd year of Huichang (841-842) in the Tang Dynasty, made his stone inscriptions when supervising the tribute tea processing on Guzhu Mountain. He wrote two poems entitled "Three Wonders in Wuxing" and "Newly Baked Tribute Tea from Huzhou".	*A Miscellany of Tea Processing* has been lost. Seen from its title, the book may have been the creation during the time of his supervision of tribute tea.

To be continued

Title	Dynasty	Author	Brief Introduction to the Author	For Reference
Book of Evaluating and Ranking	Tang	Lu Guimeng	Lu Guimeng (? -c. 881), litterateur in the Tang Dynasty, style named Luwang, with a self-styled name of "free wanderer" or "Mr. Fuli", a native of Gusu (the present Suzhou in Jiangsu Province), once an official in charge of Huzhou affairs. According to the 801st Volume of *Full Collective Works of the Tang Dynasty*, "The minister was fond of tea and located his tea garden on Guzhu Mountain, with an annual tea rental income of about 10, so meager a revenue as only to provide for his tea drinking. He wrote *Book of Evaluating and Ranking* preceded by *The Classic of Tea* and *Rhymed Formulas on Tea.*"	Also known as *Tea Savoring*, nonexistent.
Tractate on Waters for Tea	Song	Ye Qingchen	Ye Qingchen (1000-1049), style named Daoqing, from Changzhou, Suzhou, with the title of Bachelor of the Imperial Academician Courtyard and the rank of an official of the Central Ministry. As mentioned in the book, "Listed on the top in Wuxing is Zisun Tea."	One-volume book, still in existence.
Comprehensive Work on the Tea Mountain	Song	Zhu Fushi	No known records can be found about the birth or death of Zhu Fushi.	Six-volume book, already lost.

To be continued

Title	Dynasty	Author	Brief Introduction to the Author	For Reference
Manuscripts on the Tea Mountain	Yuan	Shen Zhen	Shen Zhen (c. 1363-?), style named Yuanji, with a style name of Old Man in Tea Mountain, a native of Changxing, Huzhou. At the end of the Yuan Dynasty, he was secluding in Hengyu Mountain, contented to lead a humble but virtuous life, keeping his own company, coming and going alone and chanting all the way freely. With no intention to pursue an official career, he wrote 12 volumes of *Manuscripts on the Tea Mountain*.	The book had a total of 12 volumes, but all are nonexistent. In 1738, the 3rd year of Qianlong Period of the Qing Dynasty, Bao Zhen, the then Changxing Magistrate, gathered up over a hundred poems by Shen Zhen, wrote a prelude to the collection and printed it under the new title of *Collected Remnant Works of the Old Man in the Tea Mountain* with a total of two volumes, still in existence.
Tea Picking Episodes on Guzhu Mountain	Ming	Xiao Xun	Xiao Xun was a native of Jishui, Jiangxi. In 1373, the 6th year of Hongwu Period of the Ming Dynasty, formerly a chief in the Ministry of Works, he took the new post of Changxing County Magistrate and went into Guzhu Mountain for tribute tea supervision. In 1375, the 8th year of Hongwu Period, he wrote articles asserting that those tea anecdotes were simply impossible which were claimed to have happened in the years of Dingyou (the 15th year of Yongle, ie, 1418) and Bingwu (the 1st year of Xuande, i. e., 1426) in the Ming Dynasty. It is yet to be verified.	Still existent.

To be continued

Title	Dynasty	Author	Brief Introduction to the Author	For Reference
Elucidation on Tea	Ming	Xu Cishu	Xu Cishu (1549-1604), with the given name of Ranming and a style name of Nanhua, lame but highly literate, was fond of collecting rare stones and sampling famous tea. He was productive of many works like poetry and essays. He made an intimate friend of Yao Shaoxian from Wuxing, who had a tea garden opened up in Mingyue Gorge of Guzhu Mountain. Every year when the fresh tea leaves began to sprout, the two of them would draw from the Jinsha spring and sip tea while evaluating the taste. Yao "at usual times knows the ropes of tea and imparts them all" to Xu, who, combining his personal practice and experiences from tasting the tea and the spring water, wrote the book of *Elucidation on Tea* in 1597, the 25th year of Wanli Period of the Ming Dynasty.	Still in existence.
Description of Luojie Tea	Ming	Xiong Mingyu	Xiong Mingyu (c. 1579-1649), with a style name of Liangru, a native of Jinxian County, Jiangxi, in 1601, the 29th year of Wanli Period of the Ming Dynasty, passed the provincial civil service examination. First as the Changxing Magistrate, then he was promoted to the post of Senior Supervising Secretary.	*Description of Luojie Tea*, also titled *Tea Anecdotes of Jieshan Mountain*, still existent.

To be continued

Title	Dynasty	Author	Brief Introduction to the Author	For Reference
			He bluntly exposed others' misdeeds and submitted memorials of social abuses to the emperor. His remarks caused him crises; he was involved in the Donglin Party controversy, and went through ups and downs in his officialdom, with the highest office of Minister of the Board of War before his retirement. During his administration as Changxing Magistrate, "Riding alone, he inquired about the local customs and conventions, encouraging mulberry planting and tea producing." (In 1608, the 36th year of Wanli Period of the Ming Dynasty, he wrote *Description of Luojie Tea* and "A Return Poem to Changxing Monk for His Luojie Tea" with a prelude to it.)	Still in existence.
Appraisal on Dongshan Jiecha Tea	Ming	Zhou Gaoqi	Zhou Gaoqi (? - 1645), style named Bogao, was a native from Jiangyin in Jiangsu. He was brilliant and flexible, erudite and extensively informed, expert at ancient literature. In 1637, the 10th year of Chongzhen Period of the Ming Dynasty, he was Deputy Governor of Changxing, Huzhou. The book of *Appraisal on Dongshan Jiecha Tea* was written in 1640, the 13th year of Chongzhen Period.	*Appraisal on Dongshan Jiecha Tea*, also known as *Appraisal on Dongshan Tea*, one volume, still existent.

To be continued

Title	Dynasty	Author	Brief Introduction to the Author	For Reference
Guidelines on Jiecha Tea	Ming	Feng Kebin	Feng Kebin, style named Zhengqing, was a native of Yidu (present Qingshan), Shangdong. In 1622, the 2nd year of Tianqi Period of the Ming Dynasty, he succeeded in the highest imperial examination and became a military officer in Huzhou, famous for his paintings of bamboo and rocks. He named his residence "Shipu Study", doing some writing and painting in his spare time. Mingxia, the servant girl and his concubine, was versatile and highly talented, often waiting by the writing table serving tea and dissolving the ink stick. Feng Kebin wept mournfully over the death of Mingxia and buried her in Xianshan Mountain. The poem goes as the following: "My soul even sheds tears lingering over this mountain. My dream of thee would turn all the clouds into sorrowful rain." So as it was rightfully said, he was "deeply infatuated with his affection". He went into seclusion when the Qing Dynasty started and refused to enter into officialdom. In 1624, the 4th year of Tianqi Period, he was Acting Changxing Magistrate. The book of *Guidelines on Jiecha Tea* was written in 1642, the 15th year of Chongzhen Period.	Still in existence.

To be continued

Title	Dynasty	Author	Brief Introduction to the Author	For Reference
Exclusive Studies of Jiecha Tea	Ming	Zhou Qingshu	Zhou Qingshu, the years of whose birth and death are unclear, secluded in Changxing.	Shen Zhou, calligrapher and painter in the Ming Dynasty, wrote a book entitled *Reflections on Exclusive Studies of Jiecha Tea*, narrating that "Qingshu secluded in Changxing and took tea utensils with him wherever he went. He invited me to have tea and appreciate the yellow leaves in the white tea bowls. It was a pity we could no longer see Hongjian and Junmo. So we mournfully poured two cups of tea on the ground to honor them." It is not clear whether the whole book is still in existence.
Collected Works on Jiecha Tea	Qing	Mao Xiang	Mao Xiang (1611-1693), with a style name of Bijiang and nickname of Chaomin, began to enjoy a high reputation for his virtue and capability in his early youth. At the end of the Ming Dynasty he was admitted into the Imperial College. Shi Kefa, the general, recommended him to be inspector in the army but he declined. When the Qing Dynasty began he did not take any official post, only entertaining himself through writing. Accompanied by guests and followers, he had banquets and enjoyed great popularity, hailed as one of the "Four Celebrities in the Early Qing Dynasty". His romantic stories with Dong Xiaowan the famous blonde were widely known. The book gives an account of some life anecdotes of the unequaled beauty Dong Xiaowan, including her love for Jiecha Tea.	*Collected Works on Jiecha Tea* was written at the beginning of the Qing Dynasty, in one volume, and still in existence. Around half of its contents were excerpted from books of Jiecha Tea written by Xiong Mingyu, Zhou Gaoqi and Feng Kebin.

(Compiled by Ding Kexing)

Reference

[1] Lu Yu. *The Classic of Tea* [M]. Beijing: China Workers Press, 2003.

[2] Zhuang Wanfang. *Collection of Essays on the Chinese Tea History* [M]. Beijing: Science Press, 1989.

[3] Guo Mengliang. *Chinese Tea History* [M]. Taiyuan: Shanxi Ancient Books Publishing House, 2003.

[4] Lu Yu Tea Culture Research Association of Huzhou. *Studies on Tea Culture* of Lu Yu[Z]. 1990 -2005.

[5] Ding Kexing, Zhu Wen, Wu Jianying. *Huzhou: The Cradle of Chinese Tea Culture*[Z]. 2005.

[6] Wang Zhenheng, Wang Guangzhi. *Annals of Famous Chinese Tea*[M]. Beijing: China Agriculture Press, 2000.

[7] *Zhejiang Economic Yearbook* [M]. Southeast Statistic Service, 1948.

[8] Wu Juenong. *A Commentary on The Classic of Tea* [M]. 2nd Edition. Beijing: China Agriculture Press, 2005.

[9] Editorial Committee of Records of Chinese Tea Plant Varieties. *Records of Chinese Tea Plant Varieties* [M]. Shanghai: Shanghai Scientific and Technical Publishers, 2001.

[10] Chen Zugui, Zhu Zizhen. *Selections from Historical Data of Chinese Tea*[M]. Beijing: China Agriculture Press, 1981.

[11] Ye Yu. *Integration of Tea Books: With Emendation and Annotations* [M]. Harbin: Heilongjiang People's Publishing House, 2001.

[12] Ye Yu. *A Glossary of Tea Ceremony Terms*[M]. Harbin: Heilongjiang People's Publishing House, 2002.

[13] Ye Yu. *Tea Ceremony*[M]. Harbin: Heilongjiang People's Publishing House, 2002.

[14] Ding Wen. *Chinese Tea Ceremony* [M]. Xi'an: Shaanxi Tourism Press, 1994.

[15] Qian Shilin. *Selections of Ancient Chinese Tea Poems*[M]. Hangzhou: Zhejiang Ancient Books Publishing House, 1989.

[16] Lu Yu Tea Culture Research Association of Huzhou. *Huzhou Tea Poems*[Z]. 2005.

[17] Ke Qiuxian. *Book of Tea: A Guided Reading of Tea Art, Tea Ceremony, Tea Classic and Tea Saint*[M]. Beijing: China Building Materials Press, 2003.

[18] *The Complete Works of Chinese Classics*[DB]. Electronic Version of Wenyuan Chamber Version. Shanghai: Shanghai People's Publishing House, 1999.

Postface

Before the Spring Festival last year, Mr. Shao Wei, a member of Huzhou Social Sciences Association, called and told me that the Association prepared to compile a set of Huzhou culture series. He asked me whether I was willing to compose a book on Huzhou tea culture.

Although I have done many years of research on Huzhou tea culture with senior leaders such as Dong Shuduo and Xu Mingsheng, and have published several pieces of research paper successively with the guidance of Mr. Qian Pu and Zhu Nailiang, it is a great pressure for me to compose a book on the history and present situation of Huzhou tea culture systematically. I invited many experts on Huzhou tea culture to take part in the composition of the book, only to have been declined for one reason after another. I can do nothing but ask Mr. Shao Wei to find other men to do the job.

Unexpectedly, right after the Spring Festival, Mr. Zhu Xiang, Vice President of Huzhou Social Sciences Association, called and invited me to take up the job. His invitation was too polite and decisive for me to refuse. Being so nervous, I talked about this matter with Zhong Ming, Deputy Secretary of Municipal Party Committee and Municipal Office Director. To my surprise, he agreed to compose the book with me.

For the sake of tea culture study and friendship, I shoulder bravely on the responsibility to complete the job. On my invitation, several senior scholars from Lu Yu Tea Culture Research Association of Huzhou participated in data collection in a chronological order. Each of us was responsible for one part of the job. Whenever I was free, I indulged myself in collecting the data of Huzhou tea culture and worked very hard to compose the book. A year later the book was finally finished.

In writing the book, I have cited much from the reference books and a great deal from sixteen issues of *Studies on Tea Culture of Lu Yu*. I will give my heartfelt thanks to all the authors of the above data, namely they are Dai Meng, Xu Mingde, Luo Jiaqing, Cai Yiping, Zhong Weijin, Shen Pengnian, Qian Dayu, Zhang Baoming, Qian Shilin, Xu Rongquan, Min Quan and so on. In particular, I will give my heartfelt thanks to Ding Kexing, Zhang Zhiliang and Lin Shengyou who should have enjoyed their life in their 60s or 70s, but they devoted themselves to the writing of the book. Besides, I will be grateful for the timely help from the staff of Huzhou Library who have made it so convenient for me to access to the reference materials.

Zhang Xiting

November 22nd, 2006